Ready® | 5 Mathematics INSTRUCTION

North Carolina

NCSCS Edition
Built for North Carolina's
Standard Course of Study

Vice President of Education: Adam Berkin
Associate Vice President, Custom Product & Operations: Audra Bailey
Editorial Director: Cynthia Tripp
Assistant Director, Custom Operations: Deborah Golumbek
Executive Editors: Penny Dowdy, Kathy Kellman
Senior Editor: Ruth Estabrook
Supervising Editors: Pam Halloran, Lauren Van Wart
Editors: Sarah Kraus, David Polakoff
Project Managers: Grace Izzi, Sherry Pilkerton
Cover Design and Illustrator: Matt Pollock
Book Design: Jeremy Spiegel
Composition: Edward Scanlon, Mark Nodland, Scott Hoffman

NOT FOR RESALE

ISBN 978-1-4957-7207-8
©2018–Curriculum Associates, LLC
North Billerica, MA 01862
No part of this book may be reproduced
by any means without written permission
from the publisher.
All Rights Reserved. Printed in USA.
15 14 13 12 11 10 9 8 7 6 5

Table of Contents

Mathematical Practices Handbook . . . . SMPi

Unit 1 **Number and Operations in Base Ten** 1

Lesson

1 *Understand* Place Value . . . . 2 — **NC.5.NBT.1**
2 Patterns in Products and Quotients . . . . 8 — **NC.5.NBT.1**
3 Read and Write Decimals . . . . 18 — **NC.5.NBT.3**
4 Compare and Round Decimals . . . . 28 — **NC.5.NBT.3**
5 Multiply Whole Numbers . . . . 38 — **NC.5.NBT.5**
6 Divide Whole Numbers . . . . 46 — **NC.5.NBT.6**
7 Add and Subtract Decimals . . . . 56 — **NC.5.NBT.7**
8 Multiply Decimals . . . . 68 — **NC.5.NBT.7**
9 Divide Decimals. . . . . 80 — **NC.5.NBT.7**

MATH IN ACTION Use Whole Numbers and Decimals . . . . . . 92 — NC.5.NBT.1, NC.5.NBT.3, NC.5.NBT.5, NC.5.NBT.6, NC.5.NBT.7

Interim Assessment 100

Unit 2 **Number and Operations— Fractions** 103

Lesson

10 Add and Subtract Fractions . . . . 104 — **NC.5.NF.1**
11 Add and Subtract Fractions in Word Problems . . 114 — **NC.5.NF.1**
12 Fractions as Division . . . . 124 — **NC.5.NF.3**

NCSCS Standards

Standards in boldface are the focus standards that address major lesson content.

Table of Contents continued

Unit 2 — Number and Operations—Fractions continued

Lesson

13	*Understand* Products of Fractions	134	**NC.5.NF.4**
14	Multiply Fractions Using an Area Model	140	**NC.5.NF.4**
15	*Understand* Multiplication as Scaling	150	**NC.5.NF.4**
16	Multiply Fractions in Word Problems	156	**NC.5.NF.4**
17	*Understand* Division with Unit Fractions	166	**NC.5.NF.7**
18	Divide Unit Fractions in Word Problems	172	**NC.5.NF.7**
MATH IN ACTION	Use Fractions	182	NC.5.NF.1, NC.5.NF.4, NC.5.NF.7

Interim Assessment 190

Unit 3 — Operations and Algebraic Thinking 193

Lesson

19	Evaluate and Write Expressions	194	**NC.5.OA.2**
20	Evaluate Expressions Using Properties	204	**NC.5.OA.2**
21	Analyze Patterns and Relationships	214	**NC.5.OA.3**
MATH IN ACTION	Expressions, Patterns, and Relationships	224	NC.5.OA.2, NC.5.OA.3, NC.5.NBT.7, NC.5.NF.4

Interim Assessment 232

NCSCS Standards

Standards in boldface are the focus standards that address major lesson content.

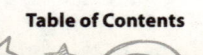

Unit 4 Measurement and Data 235

Lesson

22	Convert Measurement Units	236	**NC.5.MD.1,** NC.5.NF.7
23	Solve Word Problems Involving Conversions . . .	246	**NC.5.MD.1,** NC.5.NF.7
24	Make Line Graphs and Interpret Data	256	**NC.5.MD.2**
25	*Understand* Volume	268	**NC.5.MD.4**
26	Find Volume Using Unit Cubes	274	**NC.5.MD.4,** NC.5.MD.5
27	Find Volume Using Formulas	282	**NC.5.MD.5**
28	Find Volume of Composite Figures.	290	**NC.5.MD.5**
MATH IN ACTION	Work with Measurement and Data.	298	**NC.5.MD.1, NC.5.MD.4, NC.5.MD.5, NC.5.NBT.5**

Interim Assessment 306

Unit 5 Geometry 309

Lesson

29	*Understand* the Coordinate Plane	310	**NC.5.G.1**
30	Graph Points in the Coordinate Plane	316	**NC.5.G.1**
31	Classify Quadrilaterals.	326	**NC.5.G.3**
32	*Understand* Properties of Quadrilaterals	334	**NC.5.G.3**
MATH IN ACTION	Work with Geometry and Coordinates	340	NC.5.G.1, NC.5.G.3

Interim Assessment 348

Glossary 351

Standards, Grade 5 363

NCSCS Standards

Standards in boldface are the focus standards that address major lesson content.

©Curriculum Associates, LLC   Copying is not permitted.

MATHEMATICAL PRACTICES HANDBOOK

We use our math thinking to figure out all kinds of problems, even hard problems from real life.

There are eight math habits that will help make your math thinking grow stronger.

Keep practicing! You'll be learning to think like a math pro. Then you'll be ready to take on any problem.

THE 8 MATH HABITS

1 Solve problems.
Keep looking for clues until you solve the problem.

2 Think and reason.
Make sense of the words and the numbers in a problem.

3 Show and explain.
Share your math ideas to help others understand you.

4 Use math in the real world.
Solve problems in real life.

5 Choose a tool.
Decide when to use tools like a diagram, a ruler, or mental math.

6 Be clear and precise.
Try to be exactly right in what you say and do.

7 Zoom in and zoom out.
Look for what's the same and what's different.

8 Use patterns.
Look for patterns in math to find shortcuts.

Read more about each math habit on the pages that follow.

SMPi ©Curriculum Associates, LLC Copying is not permitted.

MATHEMATICAL PRACTICES HANDBOOK ① ② ③ ④ ⑤ ⑥ ⑦ ⑧

MATH HABIT 1

SMP 1 Make sense of problems and persevere in solving them.

Solve problems.

Keep looking for clues until you solve the problem.

For some math problems, you may not know where to start. Try different ways to find a solution and look for clues about which way works best. Then check that your answer makes sense.

To solve problems

Ask Yourself
- Can I say what the problem is asking for?
- Can I ask questions to understand it better?
- Can I think about what does or doesn't make sense?
- Can I try a different way if I need to?

Then, Discuss with a Partner
- I thought the problem didn't make sense until I asked …
- I know my answer makes sense because …

©Curriculum Associates, LLC   Copying is not permitted.

SMP1

MATHEMATICAL PRACTICES HANDBOOK 1 **2** 3 4 5 6 7 8

MATH HABIT 2

SMP 2 Reason abstractly and quantitatively.

Think and reason.

Make sense of the words and the numbers in a problem.

Reasoning is a way of thinking that puts ideas together. If you know one thing, then you know another thing. Reasoning is using math rules and common sense together.

To use reasoning to solve a problem

Ask Yourself
- Can I show how whole numbers and decimals are related?
- When I see an equation, can I think of a situation that would go with it?
- When I read a problem, can I write an equation to find the answer?
- Can I try out my answer to see if it makes sense in the problem?

Then, Discuss with a Partner
- I turned the problem into numbers when I wrote …
- I think my answer makes sense because …

SMP2

©Curriculum Associates, LLC   Copying is not permitted.

MATHEMATICAL PRACTICES HANDBOOK ① ② ③ ④ ⑤ ⑥ ⑦ ⑧

MATH HABIT 3

SMP 3 Construct viable arguments and critique the reasoning of others.

Show and explain.

Share your math ideas to help others understand you.

When you explain your math ideas to others, it helps you understand them even better. And that helps you solve other problems later. When you listen to other people, you get new ideas too.

To help explain your ideas or listen to others

Ask Yourself
- Can I use words to show how to solve the problem?
- Can I use pictures or act out the problem with objects?
- Can I ask questions to understand another person's ideas better?

Then, Discuss with a Partner
- I showed my ideas when I wrote ...
- I explained my ideas when I said ...

©Curriculum Associates, LLC   Copying is not permitted.

SMP3

MATHEMATICAL PRACTICES HANDBOOK ① ② ③ ④ ⑤ ⑥ ⑦ ⑧

MATH HABIT 4

SMP 4 Model with mathematics.

Use math in the real world.

Solve problems in real life.

One of the best ways to use your math thinking is to solve real problems. Words tell the story for the problem. Math can turn the words into a model, such as a picture or an equation.

You can use models to solve problems about shopping, art projects, sports, cooking, or … almost anything!

To solve a real-life problem

Ask Yourself
- Can I draw a picture, write an equation, or use a different model to show the math?
- Can I use my math model to solve the problem?
- Can I check that my answer makes sense?

Then, Discuss with a Partner
- I used a math model to show the problem when I …
- I know my answer makes sense because …

MATHEMATICAL PRACTICES HANDBOOK ① ② ③ ④ ⑤ ⑥ ⑦ ⑧

MATH HABIT 5

SMP 5 Use appropriate tools strategically.

Choose a tool.

Decide when to use tools like a diagram, a ruler, or mental math.

There are many tools to use in math. You can use a pencil to do a lot of math. Sometimes you need a ruler, or maybe a diagram. Often you can just do the math in your head.

To choose the best tools

Ask Yourself
- Can I do some problems in my head?
- Can I write the problem on paper?
- Can I make a table or a diagram?
- Can I use a ruler to solve the problem?

Then, Discuss with a Partner
- The tools I chose for this problem are …
- I chose these tools because …

SMP5

MATHEMATICAL PRACTICES HANDBOOK 1 2 3 4 5 **6** 7 8

MATH HABIT 6

SMP 6 Attend to precision.

Be clear and precise.

Try to be exactly right in what you say and do.

Everybody likes to be right when they do math. But sometimes people make mistakes. So it's good to check your work. And it's good to say exactly what you mean when you talk about your math ideas.

To be exactly right

Ask Yourself

- Can I use words that will help everyone understand my math ideas?
- Can I ask questions to understand the meaning of math words I don't know?
- Can I find different ways to check my work when I multiply or add?
- Can I always think about whether my answer makes sense?

Then, Discuss with a Partner

- I was careful to use the right words when I …
- I checked my answer by …

MATHEMATICAL PRACTICES HANDBOOK ① ② ③ ④ ⑤ ⑥ ⑦ ⑧

MATH HABIT 7

SMP 7 Look for and make use of structure.

Zoom in and zoom out.

Look for what's the same and what's different.

Math follows rules. Think about these equations:

$3 \times 1 = 3$

$4 \times 1 = 4$

You can *zoom out* to look at what's the *same* about problems. They show that any number times 1 is that number.

You can also *zoom in* to see what's *different* about problems. The number multiplied by 1 is different in each problem.

To zoom in and zoom out

Ask Yourself
- Can I see how decimals and fractions are both similar and different?
- Can I see how decimals and whole numbers are both similar and different?
- Can I see how shapes are different but are made from other shapes that are the same?

Then, Discuss with a Partner
- I zoomed out and used a math rule when I …
- I zoomed in and found a difference when I looked at …

©Curriculum Associates, LLC   Copying is not permitted.

SMP7

MATHEMATICAL PRACTICES HANDBOOK ① ② ③ ④ ⑤ ⑥ ⑦ ⑧

MATH HABIT 8

SMP 8 Look for and express regularity in repeated reasoning.

Use patterns.
Look for patterns in math to find shortcuts.

It's important in math to pay close attention. You might find a pattern or see a math idea.

Think about the pattern you see when you count by elevens:

11, 22, 33, 44, 55 …

You can use the pattern to make a good guess about what comes next.

To use patterns

Ask Yourself
- Can I find a pattern in a math problem?
- Can I use clear math words to describe my pattern?
- Can I make a good guess about what is next?

Then, Discuss with a Partner
- I saw a pattern in this problem when I looked at …
- I made a good guess about the pattern when I …

SMP8

©Curriculum Associates, LLC   Copying is not permitted.

Unit 1
Number and Operations in Base Ten

Let's learn about multiplying and dividing decimals.

Real-World Connection Knowing how to add, subtract, multiply, and divide whole numbers is important. But some situations do not involve whole numbers. You want to download 6 new songs that are $1.49 each. You just read that the beach you love to visit is wearing away at a rate of 17.2 centimeters each year. You have 11.5 meters of fabric and will be making 5 equal-sized table covers for the science fair.

In This Unit You will multiply and divide with whole numbers. You will also learn how to calculate with decimals. You will compare decimals, add them, subtract them, multiply them, and divide with them.

✓ Self Check

Before starting this unit, check off the skills you know below. As you complete each lesson, see how many more skills you can check off!

I can:	Before this unit	After this unit
Explain patterns in products and quotients, for example: the numbers 600,000; 60,000; 6,000; 600 show a pattern of multiplying by 0.1 or dividing by 10.	☐	☐
read and write decimals, for example: $80.63 = 8 \times 10 + 6 \times \frac{1}{10} + 3 \times \frac{1}{100}$.	☐	☐
compare decimals, for example: $3.47 > 3.096$.	☐	☐
round decimals, for example: 6.274 rounded to the nearest tenth is 6.3.	☐	☐
multiply whole numbers, for example: $410 \times 16 = 6,560$.	☐	☐
divide whole numbers, for example: $2,812 \div 38 = 74$.	☐	☐
add and subtract decimals, for example: $20.08 + 5.15 = 25.23$.	☐	☐
multiply decimals, for example: $7.25 \times 9.4 = 68.15$.	☐	☐
divide decimals, for example: $18.8 \div 4 = 4.7$.	☐	☐

©Curriculum Associates, LLC   Copying is not permitted.

Lesson 1 Introduction
Understand Place Value

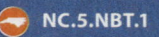

NC.5.NBT.1

Think It Through

How is place value related to the number 10?

We use a number system called **base ten**. This means that place value in any number is based on a pattern of tens.

Look at the following place-value models for whole numbers.

Thousands	Hundreds	Tens	Ones
1 thousand is **10 times** 1 hundred	1 hundred is **10 times** 1 ten	1 ten is **10 times** 1 one	1 one

1,000 + 100 + 10 + 1 = 1,111

Think Place value in decimals is just like place value in whole numbers.

Look at the following place-value models for decimal numbers.

Ones	Tenths	Hundredths
1 whole is **10 times** 1 tenth	1 tenth is **10 times** 1 hundredth	1 hundredth

1 + 0.1 + 0.01 = 1.11

In a decimal number, a digit in one place has ten times the value it would have in the place to its right.

×10 ×10 ×10 ×10 ×10
thousands hundreds tens ones tenths hundredths

Circle all of the numbers in the equation below the chart.

2 Lesson 1 Understand Place Value

©Curriculum Associates, LLC Copying is not permitted.

> **Think** 1 hundredth is 10 times 1 thousandth.

If you break one hundredth into 10 equal parts, each part is 1 thousandth of the whole.

Look at the pattern in the chart.

×$\frac{1}{10}$	0.00**1**
×$\frac{1}{10}$	0.0**1**
×$\frac{1}{10}$	0.**1**
×$\frac{1}{10}$	**1**
×$\frac{1}{10}$	**1**0
×$\frac{1}{10}$	**1**00
×$\frac{1}{10}$	**1**,000
×$\frac{1}{10}$	**1**0,000

(×10 arrows going up between each row)

> When you multiply a number by 10, the product is 10 times the number.

When the digit **1** moves one place to the left, its value is 10 times what it was. When the digit **1** moves one place to the right, its value is $\frac{1}{10}$ what it was.

Look more closely at the first two numbers in the chart.

- One hundredth is 10 times one thousandth.
  $0.01 = 10 \times 0.001$

- One thousandth is one tenth of one hundredth.
  $0.001 = 0.01 \div 10$

To write the fraction one thousandth as a decimal, write a zero in the tenths and hundredths places and a 1 in the thousandths place to the right of the decimal point.

$\frac{1}{1,000} = 0.001$

▶ Reflect

1 In which number does the digit 5 have a greater value, 0.05 or 0.005? How many times as great is the value? Explain how you know.

_____

_____

_____

Lesson 1 **Understand** Place Value

Lesson 1 Guided Instruction

Think About Place-Value Patterns

🔍 **Let's Explore the Idea** Let's explore place-value patterns with another example using models. Each grid represents 1 whole.

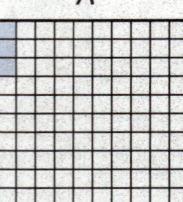

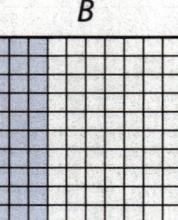

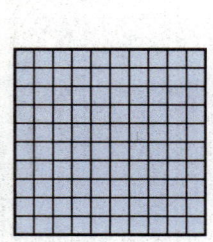

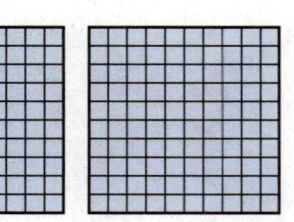

A   B   C

_____   _____   _____

2. Label models A, B, and C with a decimal to name the amount shaded.

3. The shaded region of Model B is how many times the shaded region of Model A?

   _____

   The shaded region of Model C is how many times the shaded region of Model B?

   _____

4. Use the models above to complete the equations.

   $0.03 \times 10 =$ _____

   $0.3 \times 10 =$ _____

   $0.3 \div 10 =$ _____

   $3.0 \div 10 =$ _____

Now try these two problems.

5. Continue the ×10 pattern to fill in the blanks.

   0.003   0.03   0.3   _____   _____   300

6. Use the ÷10 pattern to fill in the blanks.

   500   _____   5   0.5   0.05   _____

4 Lesson 1 Understand Place Value

Let's Talk About It

Now let's explore how 0.30 compares to 0.3.

7 Look at the models in problem 2 and the labels you wrote. Now shade some or all of the grids in the model below to show 30 hundredths. Each grid represents 1 whole.

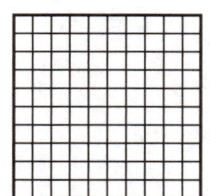

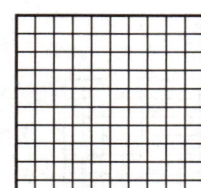

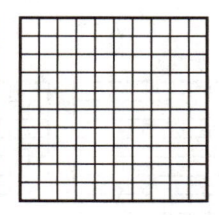

Find the model on the previous page that has the same shading as the model you shaded above. What label did you give that model on the previous page?

_____

8 Use the model you shaded in problem 7 to write a decimal for 30 hundredths.

thirty hundredths = _____

9 How do both shaded models show that 30 hundredths is the same as 3 tenths?

_____
_____

10 What equation can you write to represent "10 times 3 hundredths is 30 hundredths"?

_____

Try It Another Way

11 Imagine a model shaded to show 0.001. How much would be shaded?

_____
_____

12 How would you shade a model to show ten thousandths?

_____
_____

Lesson 1   Guided Practice

Connect   Ideas About Place-Value Patterns

Talk through these problems as a class, then write your answers below.

13 Create Shade the models below to show how the value of 0.04 is related to the value of 0.4. Then write a division equation to represent the relationship.

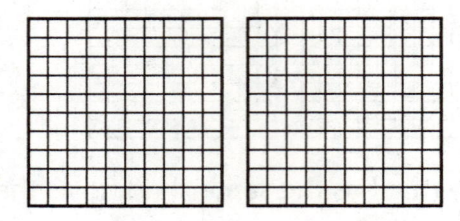

14 Analyze Kiran showed 0.08 with the model below.

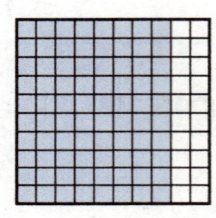

What is wrong with Kiran's model? What can be done to her model to show 0.08?

_____
_____
_____
_____

15 Demonstrate A meter is one thousandth of a kilometer. Write an equation to show the relationship between 7 meters and 0.007 kilometer.

_____
_____

Lesson 1 Independent Practice

Apply Ideas About Place-Value Patterns

16 Put It Together Use what you have learned to complete this task.

> A small beetle has a mass of 0.002 kilogram. A bee has a mass of 0.02 kilogram. A small brick has a mass of 0.2 kilogram.

Part A Write each measurement in the place-value chart below.

Object	Tens	Ones	.	Tenths	Hundredths	Thousandths

Part B Shade the models to compare the mass of the bee to the mass of the brick.

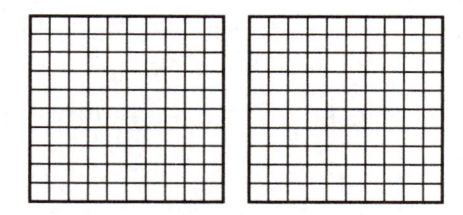

Part C Explain how many small beetles it would take to equal the mass of 1 small brick. Then write an equation using the measurements to show your answer.

_____
_____
_____
_____

Lesson 1 *Understand* Place Value   7

Lesson 2 — Introduction
Patterns in Products and Quotients

NC.5.NBT.1

Use What You Know

Take a look at this problem.

José shaded models to make a multiplication pattern. Each grid represents 1 whole. What is the next number in the pattern?

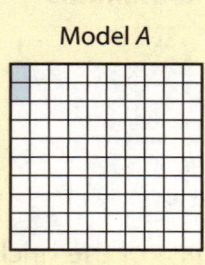

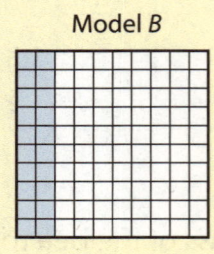

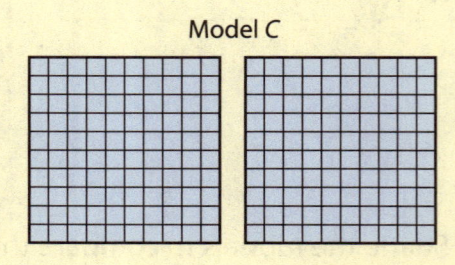

Model A   Model B   Model C

_____   _____   _____

a. What number does each model represent? Write the number below each model shown above.

b. How many times greater is the shaded part of Model B than the shaded part of Model A? _____ times

c. How many times greater is the shaded part of Model C than the shaded part of Model B? _____ times

d. Use the model above to complete the equations.

Model A	×	?	=	Model B
_____	×	_____	=	_____
Model B	×	?	=	Model C
_____	×	_____	=	_____

e. What is the pattern? Multiply by _____.

f. If the pattern continues, what would be the next number? _____

▶▶ Find Out More

You can use different strategies to explore patterns in products and quotients.

This pattern shows multiplying by 0.01: 600,000; 6,000; 60; 0.6.

Look at the pattern in a place-value chart. The digit 6 moves two places to the right from one number to the next.

Hundred Thousands	Ten Thousands	Thousands	Hundreds	Tens	Ones	.	Tenths
6	0	0	0	0	0	.	
		6	0	0	0	.	
				6	0	.	
					0	.	6

This pattern shows dividing by ten: 20; 2; 0.2; 0.02.

The place-value chart shows that the digit 2 moves one place to the right from one number to the next.

Tens	Ones	.	Tenths	Hundredths
2	0	.		
	2	.		
	0	.	2	
	0	.	0	2

▶ Reflect

1 The second chart above shows the pattern when dividing by 10 repeatedly starting at 20. Find the pattern you get from multiplying by 10 repeatedly starting at 20. How does the digit 2 move when dividing by 10 compared to how the digit 2 moves when multiplying by 10?

_____

_____

_____

Lesson 2 Patterns in Products and Quotients

Lesson 2 Modeled and Guided Instruction

Learn About: Using Multiplication Patterns

Read the problem below. Then explore different ways to describe multiplication patterns.

> Leah is playing a math game. She wrote this multiplication pattern.
>
> 0.007; 0.7; 70; 7,000
>
> What is the next number in the pattern?

▶ **Model It** You can use a place-value chart to see how the digit 7 moves from one number to the next.

Thousands	Hundreds	Tens	Ones	.	Tenths	Hundredths	Thousandths
			0	.	0	0	7
			0	.	7		
		7	0	.			
7	0	0	0	.			

▶ **Model It** You can line up the numbers on the digit 7 to see how the decimal point moves from one number to the next.

```
0.007
0.7
70.
7,000.
```

10 Lesson 2 Patterns in Products and Quotients

▶ **Connect It** Now you will use the models on the previous page to find the next number in the pattern.

To read a decimal, you tell how many of the smallest fractional part it represents.

2 Look at the first *Model It*. Write the place value of the digit 7 in each number.

0.007: There are 7 _____.

0.7: There are 7 _____.

70: There are 7 _____.

7,000: There are 7 _____.

3 How many places did the digit 7 move left from one number to the next in the pattern? _____

4 Look at the second *Model It*. How many places did the decimal point move right from one number to the next? _____

5 Every place in the place-value chart is 10 times the value of the place to its right. So every place is 10 × 10 or _____ times the value of two places to its right.

6 To find the next number in the pattern, multiply by _____.

7 What is the next number in the pattern? Explain how you found your answer.

_____
_____

8 How do you find the number that comes next in a multiplication pattern?

_____
_____
_____

▶ **Try It** Use what you just learned about multiplication patterns to solve this problem.

9 Felix multiplied a number by 10 to make a pattern that has four numbers. What could the pattern be? _____

©Curriculum Associates, LLC   Copying is not permitted.

Lesson 2 Patterns in Products and Quotients   **11**

Lesson 2 — Modeled and Guided Instruction

Learn About: Using Division Patterns

Read the problem below. Then explore different ways to describe division patterns.

> Deon wrote this division pattern.
> 200,000; 2,000; 20; 0.2
> What is the next number in the pattern?

▶ **Model It** You can use a place-value chart to see how the digit 2 moves from one number to the next.

Hundred Thousands	Ten Thousands	Thousands	Hundreds	Tens	Ones	.	Tenths
2	0	0	0	0	0	.	0
		2	0	0	0	.	
				2	0	.	
					0	.	2

▶ **Model It** You can line up the numbers on the digit 2 to see how the decimal point moves from one number to the next.

200,000.
  2,000.
     20.
      0.2

Lesson 2 Patterns in Products and Quotients

Connect It Now you will use the models on the previous page to find the next number in the pattern.

10 Look at the first *Model It*. Write the place value of the digit 2 in each number.

200,000: There are 2 _____.

2,000: There are 2 _____.

20: There are 2 _____.

0.2: There are 2 _____.

11 How did the digit 2 move from one number to the next?
_____

12 Look at the second *Model It*. How did the decimal point move from one number to the next? _____

13 To find the next number in the pattern, divide by _____.

14 What is the next number in the pattern? Explain how you found your answer.
_____
_____

15 How do you find the number that comes next in a division pattern?
_____
_____

Try It Use what you just learned about decimal patterns to solve these problems.

16 Ana divided by 100 to make a pattern. The pattern starts with 60,000. What are the next three numbers in the pattern? _____

17 Is this pattern made by dividing by 100? Explain how you know.

500,000; 50,000; 5,000; 500
_____
_____
_____

Lesson 2 Patterns in Products and Quotients

Lesson 2 Guided Practice

Practice Using Patterns in Products and Quotients

Study the example below. Then solve Problems 18–20.

Example

What do you multiply by to make this pattern?

0.003; 3; 3,000

Look at how you could use the decimal points when you line up the numbers on the digit 3.

0.003
   3.
   3,000.

The decimal point moves three places to the right from one number to the next. Each place is 10 times the place to its right. So moving three places is 10 × 10 × 10, or 1,000.

Solution _Multiply by 1,000._

How many places does the decimal point move from one number to the next?

Pair/Share
Do all the numbers follow the same pattern?

---

18 What pattern describes these numbers?

400,000; 4,000; 40; 0.4

Show your work.

Are the numbers in the pattern getting larger or smaller?

Pair/Share
Is there more than one way to solve the problem? If so, what is the other way?

Solution _____

19 What is the next number in the pattern? Describe the pattern using multiplication or division.

80; 8; 0.8, 0.08; _____

Show your work.

I need to know how the pattern was made to find the next number.

Pair/Share
How could you use the inverse operation to check your answer?

Solution _____

---

20 Which pattern was made by multiplying by 0.01?

A 900,000; 9,000; 900; 90

B 900,000; 90,000; 9,000; 900

C 900,000; 9,000; 90; 0.9

D 900,000; 9,000; 90; 9

Sam chose **D** as the correct answer. How did he get this answer?

_____

_____

_____

Remember that every number must be 0.01 times the previous number.

Pair/Share
How did you find the answer?

©Curriculum Associates, LLC   Copying is not permitted.   Lesson 2 Patterns in Products and Quotients

Lesson 2 — Independent Practice

Practice: Practice Using Patterns in Products and Quotients

Solve the problems.

1 Which statement describes the pattern? Circle the letter for all that apply.

700,000; 70,000; 7,000; 700

A Multiply by 0.1.

B Multiply by $\frac{1}{10}$.

C Multiply by 10.

D Divide by 10.

E Divide by 100.

2 Choose *Yes* or *No* to tell if the pattern was made by multiplying by 0.01.

a. 800,000; 8,000; 80; 0.08 ☐ Yes ☐ No

b. 7,000; 70; 0.7; 0.007 ☐ Yes ☐ No

c. 30; 3; 0.3; 0.03 ☐ Yes ☐ No

d. 500,000; 5,000; 50; 0.5 ☐ Yes ☐ No

3 Jason made this pattern. What is the missing number in the pattern?

6,000; 600; 60; 6; _____, 0.06

A 0.006

B 0.06

C 0.6

D 6.06

4 Which pattern shows multiplying by 1,000?

A 0.01; 100; 10,000; 1,000,000

B 1; 100; 1,000; 1,000,000;

C 0.001; 1; 1,000; 100,000

D 0.001; 1; 1,000; 1,000,000

5 Use multiplication and division to describe this pattern.

300,000; 3,000; 30; 0.3

Show your work.

Solution _____

6 Look at the pattern.

0.002; 0.2; 20; 2,000

Part A Describe the pattern using multiplication.

Show your work.

Answer _____

Part B What is the next number in the pattern? Explain how you know.

_____
_____
_____
_____

✓ **Self Check** Go back and see what you can check off on the Self Check on page 1.

Lesson 2 Patterns in Products and Quotients

Lesson 3 Introduction
Read and Write Decimals

NC.5.NBT.3

🅖 Use What You Know

You already know how to read and write whole numbers and fractions. Reading and writing decimals is similar.

> Jessica is reading aloud from a science website. She comes across the measurements 0.32 meter and 0.543 meter. How should she read 0.32?

Jessica can read 0.32 as 3 tenths and 2 hundredths, but it is easier to understand her if she says how many hundredths that is.

To read 0.32 as a number of hundredths:

a. Write 3 tenths as a fraction. _____

b. Rewrite 3 tenths as an equivalent fraction with a denominator of 100. _____

c. Write 2 hundredths as a fraction. _____

d. Complete the equation to add the like fractions for 3 tenths and 2 hundredths.

$$\frac{\Box}{100} + \frac{\Box}{100} = \frac{\Box}{\Box}$$

e. Write in words how Jessica should read the number 0.32. _____
_____

f. Explain how you can use place value and fractions to read a number with two digits to the right of the decimal point.

_____
_____
_____
_____

Find Out More

You can read decimals as strings of digits, for example: *zero point three two*. But to give meaning to the amount a decimal represents, you name the place value of the smallest-sized unit and read the number to say how many of those units.

Ones	.	Tenths	Hundredths	Thousandths
0	.	3	2	

The least place value in 0.32 is hundredths, so you read the decimal by saying how many hundredths there are. Read 0.32 as *thirty-two hundredths*.

Now look at the same decimal with a zero in the thousandths place.

Ones	.	Tenths	Hundredths	Thousandths
0	.	3	2	0

In the decimal 0.320, the least place value is thousandths. Even though 0.320 is equal to 0.32, you read 0.320 to tell how many thousandths there are. Read 0.320 as *three hundred twenty thousandths*.

The chart below shows the decimal 0.543.

Ones	.	Tenths	Hundredths	Thousandths
0	.	5	4	3

The least place value in 0.543 is thousandths. You read the decimal to tell how many thousandths there are. Read 0.543 as *five hundred forty-three thousandths*.

When a decimal number includes a whole number, read it the way you read mixed numbers. You read the decimal point as *and*. You read $3\frac{5}{10}$ as *three and five tenths*. Read 3.5 as *three and five tenths*.

▶ Reflect

1 Write 1.005 as a mixed number. Then write how to read this number aloud.

___

___

Lesson 3 Read and Write Decimals **19**

Lesson 3 Modeled and Guided Instruction

Learn About Reading a Decimal

Read the problem below. Then explore different ways to represent decimals.

> Josh has been tracking the growth of his dog since it was a puppy. He uses a meter stick to measure the dog's height and records the height as 0.604 meter. Josh's mom asks, "How tall is your dog?" What does Josh tell his mom?

▶ **Model It** You can use place-value understanding to write the expanded form of 0.604. You can also write it as a fraction.

With decimals:

$$0.604 = 0.6 + 0.004$$
$$= 6 \times 0.1 + 4 \times 0.001$$

With fractions:

$$0.604 = 6 \times \frac{1}{10} + 4 \times \frac{1}{1,000}$$
$$= \frac{6}{10} + \frac{4}{1,000}$$
$$= \frac{600}{1,000} + \frac{4}{1,000}$$
$$= \frac{604}{1,000}$$

▶ **Model It** You can write 0.604 in a place-value chart to show the place value of each digit.

Ones	.	Tenths	Hundredths	Thousandths
0	.	6	0	4

The least place value of 0.604 is thousandths.

20 Lesson 3 Read and Write Decimals

▶ **Connect It** Now you will write the word form of the decimal on the previous page using both models.

To read a decimal, you tell how many of the smallest fractional part it represents.

2 Look at the place-value chart. What are the names of the four place values in 0.604?

_____

Which of those places has the least value? _____

3 Look at the expanded form with fractions. How many thousandths are there altogether in 0.604? _____

4 What is the word form of 0.604?

_____

5 What is the word form of 1.604?

_____

6 Explain how reading the digits to the right of the decimal point and knowing the name of the least place value help you read a decimal number. Use the examples 0.604 and 1.604 in your explanation.

_____

_____

_____

_____

_____

_____

▶ **Try It** Use what you just learned about identifying the least place value to read decimal numbers. Show your work on a separate sheet of paper.

7 What is the word form of 0.44? _____

8 What is the word form of 1.057? _____

Lesson 3 Read and Write Decimals **21**

Lesson 3 Modeled and Guided Instruction

Learn About: Writing a Mixed Number as a Decimal

Read the problem below. Then explore different ways to think about writing the numbers as decimals.

> A school held a running race.
> - Abha finished the race one and sixteen thousandths seconds ahead of Nadia.
> - Brianne finished two and thirty-five hundredths seconds ahead of Chandra.
>
> What decimals represent these measurements?

▶ **Model It** Model the measurements with mixed numbers and expanded form.

Write mixed numbers to show the number of whole seconds and the fraction of a second given by each measurement. Then expand.

one *and* sixteen thousandths

$1\frac{16}{1,000}$

$1 + \frac{16}{1,000}$

$1 + \frac{10}{1,000} + \frac{6}{1,000}$

$1 + \frac{1}{100} + \frac{6}{1,000}$

two *and* thirty-five hundredths

$2\frac{35}{100}$

$2 + \frac{35}{100}$

$2 + \frac{30}{100} + \frac{5}{100}$

$2 + \frac{3}{10} + \frac{5}{100}$

▶ **Model It** Use a place-value chart to write the measurements.

Ones	.	Tenths	Hundredths	Thousandths
1	.	0	1	6
2	.	3	5	

16 thousandths is
10 thousandths and 6 thousandths or
1 hundredth and 6 thousandths.

35 hundredths is
30 hundredths and 5 hundredths or
3 tenths and 5 hundredths.

Lesson 3 Read and Write Decimals

> **Connect It** Now you will use the models to write the numbers from the problem on the previous page as standard decimals.

9 Compare the mixed-number and the place-value representations of *one and sixteen thousandths*. How does each model show the number of whole seconds?

_____

_____

10 How does each representation of *one and sixteen thousandths* show the value of the fractional part of a second?

_____

_____

_____

11 How can you write *one and sixteen thousandths* as a decimal? _____

12 How does each representation of *two and thirty-five hundredths* show the value of the fractional part of a second?

_____

_____

_____

13 How can you write *two and thirty-five hundredths* as a decimal? _____

14 Explain how you can write the standard form of a decimal given in word form.

_____

_____

_____

> **Try It** Use what you just learned to write numbers as decimals.

15 six and fifty-four thousandths _____

16 ten and seventy-five hundredths _____

Lesson 3 Read and Write Decimals   23

Lesson 3 — Guided Practice

Practice: Reading and Writing Decimals

Study the example below. Then solve problems 17–19.

Example

The height of Coach Roberts is two and fifty-four thousandths meters. Write this height as a decimal.

Look at how you could show your work using expanded form.

Two and fifty-four thousandths is $2\frac{54}{1{,}000}$.

$$\frac{54}{1{,}000} = \frac{50}{1{,}000} + \frac{4}{1{,}000}$$
$$= 5 \times \frac{10}{1{,}000} + 4 \times \frac{1}{1{,}000}$$
$$= 5 \times \frac{1}{100} + 4 \times \frac{1}{1{,}000}$$
$$= 5 \times 0.01 + 4 \times 0.001$$

Ones	.	Tenths	Hundredths	Thousandths
2	.	0	5	4

Solution _2.054 meters_

What place-value relationship lets me write $\frac{50}{1{,}000}$ as $\frac{5}{100}$?

Pair/Share
Why can you use fractions to represent a decimal number?

---

17 Aubrey runs the 100-meter dash in 14.895 seconds. What words could you use to read 14.895?

Show your work.

How can you write 14.895 as a mixed number?

Pair/Share
Is there a different way to use words to represent 14.895?

Solution _____

24   Lesson 3 Read and Write Decimals

18 Paulo measured the width of a sideline on a football field. It was one hundred two thousandths of a meter. Write this width as a fraction and as a decimal in expanded form.

Show your work.

How could you represent this number in a place-value chart?

Solution _____

_____

Pair/Share
What other ways could you represent this number?

---

19 Which of the following is a representation of 4.082? Circle the letter of the correct answer.

A $4 + \frac{8}{10} + \frac{2}{1,000}$

B four and eighty-two thousandths

C four and eight and two hundredths

D $4 + 8 \times \frac{1}{10} + 2 \times \frac{1}{100}$

What are the place values of each digit in 4.082?

Rachel chose **D** as the correct answer. How did she get that answer?

_____

_____

_____

Pair/Share
Does Rachel's answer make sense?

Lesson 3 — Independent Practice

Practice: Reading and Writing Decimals

Solve the problems.

1 Daniel ran the 400-meter dash in 89.023 seconds. Which of the following expresses this time in words?

   A eighty-nine and twenty-three hundredths seconds

   B eighty-nine and two tenths and three thousandths seconds

   C eighty-nine and twenty-three thousandths seconds

   D eighty-nine thousand twenty three seconds

2 What decimal represents $6 \times 1{,}000 + 2 \times 10 + 3 \times \frac{1}{10} + 5 \times \frac{1}{1{,}000}$?

   A 6,020.305

   B 6,200.350

   C 6,020.035

   D 6,002.035

3 A guitar string that plays a very high note is eleven thousandths of an inch thick. A bass string that plays a very low note is ten times as thick. For each quantity, choose either *Yes* or *No* to tell whether it is equal to ten times eleven thousandths.

   a. $\frac{1}{100} + \frac{1}{1{,}000}$ ☐ Yes ☐ No

   b. $\frac{1}{10} + \frac{1}{100}$ ☐ Yes ☐ No

   c. 0.11 ☐ Yes ☐ No

   d. eleven hundredths ☐ Yes ☐ No

   e. eleven ten thousandths ☐ Yes ☐ No

4 Which of the following correctly represent 57.036? Circle the letter for all that apply.

A $57 + \frac{3}{100} + \frac{6}{1,000}$

B $57 + 3 \times \frac{10}{1,000} + 6 \times \frac{10}{1,000}$

C $57 + 36 \times 0.01$

D $57 + 36 \times 0.001$

E fifty-seven and thirty-six hundredths

5 Represent 240.149 in two different ways. Then explain how each way shows the place value of the digits of the number.

_____
_____
_____
_____
_____
_____
_____
_____

6 Alex wrote 103.903 in expanded form as $100 + 3 \times 1 + 9 \times \frac{1}{100} + 3 \times \frac{1}{1,000}$. Explain his mistake. Then tell how to correct it.

_____
_____
_____
_____

✓ Self Check Go back and see what you can check off on the Self Check on page 1.

Lesson 4   Introduction
Compare and Round Decimals

Use What You Know

You already know how to compare and round whole numbers. In this lesson, you will use place-value understanding to compare and round decimals.

> Grace has a collection of buttons in a jar. The blue buttons are 0.008 meter wide and the red buttons are 0.02 meter wide. Which color button is wider?

a. In the numbers 0.008 and 0.02, what are the place values of the 8 and the 2?

8 is in the _____ place.

2 is in the _____ place.

b. Compare those place values: There are _____ thousandths in 1 hundredth.

c. How do you know that 8 thousandths is less than 1 hundredth?
_____

d. Write an inequality statement using < or > to compare 0.008 and 0.01. _____

e. Now think about whether 0.02 is greater or less than 0.01. Then write an inequality statement to compare 0.008 and 0.02. _____

f. Explain how you know this inequality statement is correct.

_____
_____
_____

g. Which color button is wider? _____

28   Lesson 4 Compare and Round Decimals

▶▶ Find Out More

To compare decimals, you can think about how to express numbers in different ways. Each row in the table shows other ways to express the same number.

Words	Place Value	Fractions
10 hundredths is 1 tenth	0.10 = 0.1	$\frac{10}{100} = \frac{1}{10}$
10 thousandths is 1 hundredth	0.010 = 0.01	$\frac{10}{1,000} = \frac{1}{100}$
100 thousandths is 1 tenth	0.100 = 0.1	$\frac{100}{1,000} = \frac{1}{10}$

The decimal 0.008 is 8 thousandths. Since 8 thousandths is less than 10 thousandths, 0.008 < 0.010.

Expressing numbers in different ways can also help you round decimals. To round 0.008 to the nearest hundredth is to determine whether it is closer to 0.00 or 0.01.

Just like you do with a whole number, you **round a decimal down** when it is **less than halfway** between two values. You always **round a decimal up** when it is **more than halfway, or exactly halfway**, between the two values.

But, what *is* halfway between 0 and 0.01, or halfway between 0.01 and 0.02?

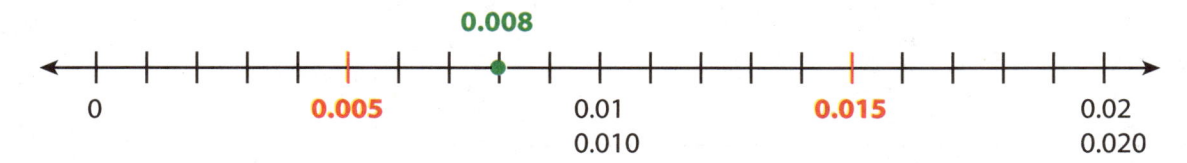

Since half of 10 thousandths is 5 thousandths, halfway between 0 and 0.010 is **0.005**. Since 0.008 is more than halfway from 0 to 0.01, round 0.008 up to 0.01.

▶ Reflect

1 Look at the number line above. Is 0.013 more or less than halfway between 0.01 and 0.02? How do you know?

_____

_____

_____

Lesson 4 Compare and Round Decimals

Lesson 4 **Modeled and Guided Instruction**

Learn About > **Comparing Decimals**

Read the problem. Then explore different ways to compare decimals.

> Gary and Melissa compare the distances they walk each morning to get to school.
> - Gary walks 3.275 kilometers.
> - Melissa walks 3.24 kilometers.
>
> Who walks farther to get to school?

▶ **Model It** Express the two distances as mixed numbers with like denominators.

Gary: $3.275 = 3\frac{275}{1,000}$ Melissa: $3.24 = 3\frac{24}{100}$

$3\frac{24}{100} = 3\frac{240}{1,000}$

$3\frac{275}{1,000} > 3\frac{240}{1,000}$, so Gary walks farther to get to school.

▶ **Model It** Write the distances in a place-value chart.

	Ones	.	Tenths	Hundredths	Thousandths
Gary	3	.	2	7	5
Melissa	3	.	2	4	0

Gary walks 3,275 thousandths kilometers to get to school.

Melissa walks 3,240 thousandths kilometers to get to school.

3,275 thousandths > 3,240 thousandths, so Gary walks farther to get to school.

Lesson 4 Compare and Round Decimals   ©Curriculum Associates, LLC   Copying is not permitted.

Connect It Now you will use the two models on the previous page to explore the same problem. The first *Model It* shows $3\frac{24}{100}$ as $3\frac{240}{1,000}$. The second *Model It* shows 3.24 as 3.240.

2. Both *Model Its* change the way the fractional part or decimal part of 3.24 is represented. What is the same about how the denominator of the fraction and the place value of the decimal change? _____

3. What is the same about how the numerator of the fraction and the digits to the right of the decimal point change? _____

4. How does rewriting the fractional part or decimal part of 3.24 make it easier to compare 3.24 to 3.275? _____

5. Explain how you can use what you know about comparing 240 and 275 to help you compare $3\frac{240}{1,000}$ with $3\frac{275}{1,000}$ or compare 3.240 with 3.275. _____

Try It Use what you just learned about comparing decimals to solve these problems. Show your work on a separate sheet of paper.

6. Heather's rabbit has a mass of 5.190 kilograms. Jeff's rabbit has a mass of 5.195 kilograms. Write an inequality statement comparing the masses of the rabbits. _____

7. Brayden and his cousin Kayla live twenty miles away from each other. On the same day, Brayden recorded 1.046 inches of rain at his house and Kayla recorded 1.062 inches of rain at her house. Whose house received fewer inches of rain that day? _____

Lesson 4 Compare and Round Decimals

Lesson 4 Modeled and Guided Instruction

Learn About Rounding Decimals

Read the problem. Then explore different ways to round the decimals.

Maura has a shelf that is 2.97 meters long. She is buying craft boxes that are 0.273 meter long. Maura wants to round these numbers to the nearest tenth to estimate how many boxes will fit on the shelf. About how many boxes will fit on the shelf?

▶ **Model It** Place 0.273 on a number line to see its relationship to nearby tenths.

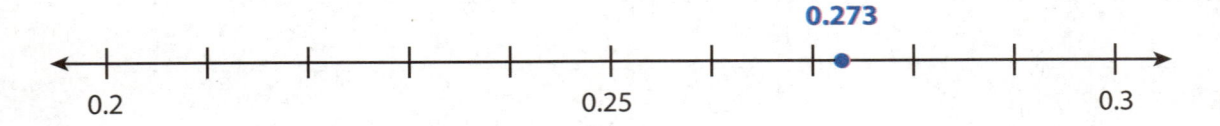

0.273 is more than halfway from 0.2 to 0.3. It is closer to 0.3.

▶ **Model It** Use a place-value chart to compare 2.97 to nearby tenths.

Ones	.	Tenths	Hundredths
2	.	9	0
2	.	9	1
2	.	9	2
2	.	9	3
2	.	9	4
2	.	9	5
2	.	9	6
2	.	9	7
2	.	9	8
2	.	9	9
3	.	0	0

2.97 is more than halfway between 2.9 and 3.0, so it rounds up to 3.0.

▶ **Connect It** Now compare the two ways to think about rounding numbers.

8 On the number-line model, how is the halfway point between the two nearby tenths shown? _____
_____

9 In the chart, how is the halfway point between the two nearby tenths shown?
_____

10 To the nearest tenth, how long is the shelf? _____

11 To the nearest tenth, how long is a craft box? _____

12 About how many boxes will fit on the shelf? How do you know?
_____

13 You can use the digit to the right of the place you want to round to tell if you should round up or down. Explain how.
_____
_____

14 Describe how each model can help you round a decimal. _____
_____
_____

15 When a number is exactly halfway between nearby tenths, you round it up. Round 0.25 to the nearest tenth. _____

▶ **Try It** Use what you just learned about rounding decimals to solve these problems. Show your work on a separate sheet of paper.

16 A dropper holds 0.813 milliliter of fluid. What is this amount rounded to the nearest hundredth? _____

17 At the end of each month, Belinda earns $0.024 for every dollar in her savings account. Round this to the nearest hundredth to show about how many cents she earns for every dollar. _____

Lesson 4 Compare and Round Decimals

Lesson 4   Guided Practice

Practice > Comparing and Rounding Decimals

Study the example below. Then solve problems 18–20.

Example

Makayla's mass at birth was 3.747 kilograms. Her brother John's mass at birth was 3.477 kilograms. Whose mass at birth rounded to the nearest whole kilogram was 4 kilograms?

Look at how you could show your work using a number line.

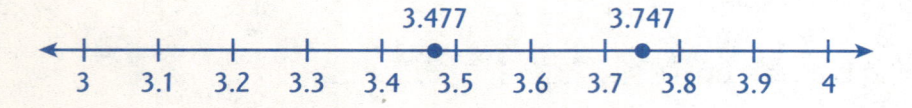

```
                    3.477       3.747
   ←——+——+——+——+——+——●——+——+——●——+——+——→
   3  3.1 3.2 3.3 3.4 3.5 3.6 3.7 3.8 3.9  4
```

The point representing Makayla's mass at birth is closer to 4 kilograms than to 3 kilograms. The point representing John's mass at birth is closer to 3 kilograms than to 4.

Solution _Makayla's mass at birth rounds to 4 kilograms._

Rounding decimals is like rounding whole numbers.

Pair/Share
How do the masses at birth compare?

18 A grocery store charges to the nearest hundredth of a pound for produce. If Landon purchased 2.386 pounds of tomatoes, what weight did the store charge for?

Show your work.

How do I use place value to round whole numbers and decimals?

Pair/Share
About how many whole pounds of tomatoes did Landon purchase? About how many tenths of a pound did he purchase?

Solution _____

Lesson 4 Compare and Round Decimals

19 Ellie fills her car with 14.297 gallons of gas. Alivia fills her car with 14.209 gallons of gas. Write an inequality statement comparing the two gas purchases.

Show your work.

How do I use place value to compare numbers?

Pair/Share
How could you use rounding to compare the numbers?

Solution _____

20 Which of the following comparisons is NOT true? Circle the letter of the correct answer.

A 42.300 = 42.3

B 5.020 < 5.20

C 0.149 < 1.490

D 108.308 > 108.4

Ava chose **A** as the correct answer. How did she get that answer?

_____

_____

_____

_____

Do I compare the number of digits or compare the place value of the digits?

Pair/Share
Does Ava's answer make sense?

Lesson 4 Independent Practice

Practice: Comparing and Rounding Decimals

Solve the problems.

1. The weights of four similar packs of tomatoes are listed below.

   Pack A: 2.456 pounds
   Pack B: 2.457 pounds
   Pack C: 2.454 pounds
   Pack D: 2.459 pounds

   Malcolm rounds the weights to the nearest hundredth pound. Which weight does NOT round to 2.46 pounds?

   A 2.456 pounds

   B 2.457 pounds

   C 2.454 pounds

   D 2.459 pounds

2. Which comparison is NOT true?

   A $1.78 < 1.92$

   B $1\frac{78}{100} < 1\frac{92}{100}$

   C $1 + 7 \times \frac{1}{10} + 8 \times \frac{1}{100} < 1 + 9 \times \frac{1}{10} + 2 \times \frac{1}{100}$

   D $1 + \frac{70}{10} + \frac{80}{100} < 1 + \frac{9}{10} + \frac{2}{100}$

3. Tori and Alvin use a stopwatch to record the amount of time it takes a marble to roll down a four-story ramp. Tori's marble takes 2.845 seconds, while Alvin's marble takes 2.835 seconds. Tell whether each sentence is *True* or *False*.

   a. Rounded to the nearest tenth, both times are the same.   ☐ True ☐ False

   b. Rounded to the nearest hundredth, both times are the same.   ☐ True ☐ False

   c. Both times are closer to 2.8 than 2.9.   ☐ True ☐ False

   d. Both times are more than halfway between 2.84 and 2.85.   ☐ True ☐ False

   e. Alvin's marble was faster.   ☐ True ☐ False

4 Lauren is going to have soup for dinner, but she doesn't want more than 0.92 gram of sodium. Decide if each brand of soup has less than, greater than, or exactly 0.92 gram of sodium. Write each brand in the correct category in the chart.

Brand A: 0.920 gram   **Brand B:** 0.875 gram   **Brand C:** 0.915 gram   **Brand D:** 1.100 gram

Less than 0.92 gram	Equal to 0.92 gram	Greater than 0.92 gram

5 The weights of four packages of meat are shown below.

Package A	1.748 pounds
Package B	1.645 pounds
Package C	1.754 pounds
Package D	1.532 pounds

Part A Each package is priced by how much it weighs to the nearest tenth of a pound. Order the packages from greatest to least weight.

Show your work.

Solution _____

Part B Explain how the order would change if you rounded the weights to the nearest hundredth.

_____

_____

_____

✓ **Self Check**   Go back and see what you can check off on the Self Check on page 1.

Lesson 5 Introduction
Multiply Whole Numbers

NC.5.NBT.5

Use What You Know

In grade 4, you multiplied two-digit numbers by two-digit numbers. Now you'll multiply three-digit numbers by two-digit numbers. Take a look at this problem.

A mall is designing an outdoor space. The available space is 127 feet by 46 feet. The plan is for a grassy section with a width of 40 feet. Next to the grass will be a cement sidewalk with a width of 6 feet. What is the area of the available space in square feet?

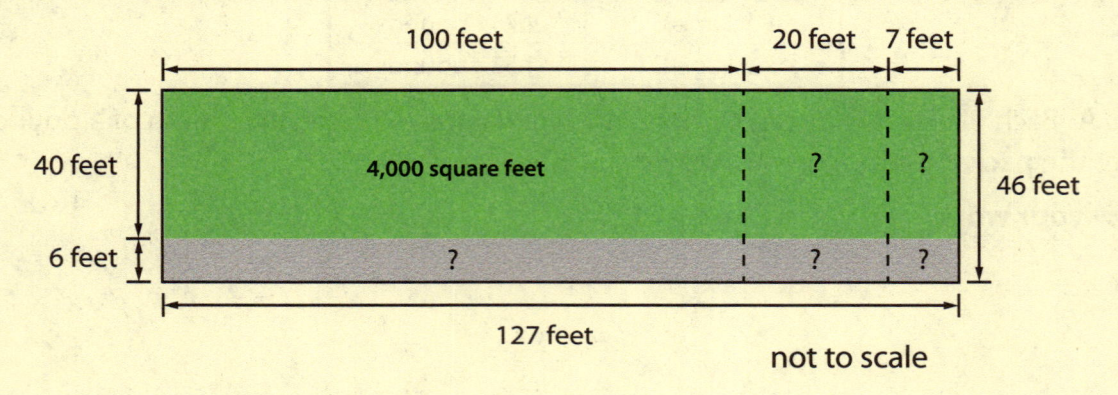

not to scale

The area model above shows the length and width of each space.

a. What is the total length of the outdoor space? _____

   What is the total width? _____

b. How is the area of each space found? _____
   _____

c. How can you find the total area? _____

d. The product of 127 and 46 is the total area of the outdoor area. How can you find the area for grass? _____

   How can you find the area for the sidewalk? _____
   _____

e. What is the total area? _____

38 Lesson 5 Multiply Whole Numbers ©Curriculum Associates, LLC   Copying is not permitted.

▶▶ Find Out More

One way to multiply multi-digit numbers is to find **partial products**. Break apart one of the factors into parts. Then multiply each of the parts by the other factor to get parts of the product, or partial products. Then, add the partial products to find the total product.

- **One way** to show using partial products to find 127 × 46:

    Break apart 46 into (40 + 6): 127 × **46** = 127 × (**40** + **6**)

    Find each partial product.

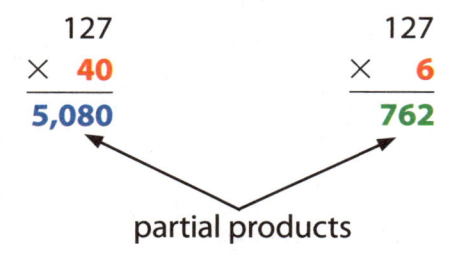

    ```
       127              127
    ×   40           ×    6
    ─────            ─────
     5,080              762
    ```

    partial products

    Then find the sum of the two partial products.

    5,080 + **762** = 5,842

- **Another way** to show using partial products to find 127 × 46:

    ```
        127
    ×    46
    ──────
        762  ← area for sidewalk
    + 5,080  ← area for grass
    ──────
      5,842  ← total area
    ```

▶ Reflect

1 How is multiplying by a three-digit number similar to multiplying by a two-digit number?

_____

_____

_____

_____

Lesson 5 Multiply Whole Numbers **39**

Lesson 5 👥 **Modeled and Guided Instruction**

Learn About > Multiplying Three-Digit Numbers

Read the problem below. Then explore different ways to multiply a three-digit number by a two-digit number.

> There are 128 pens in a full box. How many pens are in 35 full boxes?

▶ **Model It** Use an area model to show partial products.

Sketch a rectangle with dimensions 128 by 35.

128 × 35

128 is **100 + 20 + 8**.

35 is **30 + 5**.

	100	**+**	**20**	**+**	**8**
30	30 × 100 = 3,000		30 × 20 = 600		30 × 8 = 240
+ 5	5 × 100 = 500		5 × 20 = 100		5 × 8 = 40

First row: 3,000 + 600 + 240 = 3,840

Second row: 500 + 100 + 40 = 640

So, 128 × 35 = 3,840 + 640 = 4,480.

▶ **Model It** Use the distributive property to find partial products and add them.

128 × 35 = 128 × (30 + 5)

128 × (30 + 5) = (128 × 30) + (128 × 5)

```
     128                    128
   ×  30                  ×   5
   ─────                  ─────
     240 → (30 × 8)         40 → (5 × 8)
     600 → (30 × 20)       100 → (5 × 20)
 + 3,000 → (30 × 100)    + 500 → (5 × 100)
   ─────                  ─────
   3,840                    640
```

3,840 + 640 = 4,480

> **Connect It** Now you will compare the two ways to multiply using partial products.

2 Why is the area model divided into six sections? _____

_____

_____

3 How do the three partial products in each multiplication equation in the second *Model It* relate to the three sections in each row of the area model? _____

_____

_____

4 Would the product change if 30 and 5 on the left side of the area model were changed to 20, 10, and 5? Explain. _____

_____

_____

_____

_____

5 List two different ways you could break up the factors in 239 × 64 to find the product. Explain why both ways would have the same product. _____

_____

_____

_____

_____

> **Try It** Use what you just learned about multiplying numbers to solve this problem. Show your work on a separate sheet of paper.

6 A bookshelf at a library holds 156 books. There are 15 bookshelves in the children's section. How many children's books can the library place on the bookshelves?

_____

©Curriculum Associates, LLC   Copying is not permitted.

Lesson 5 Multiply Whole Numbers   **41**

Lesson 5 Guided Practice

Practice: Multiplying Whole Numbers

Study the example below. Then solve problems 7–9.

Example

There are 366 days in a leap year and 24 hours in a day. How many hours are in a leap year?

Look at how you could show your work.

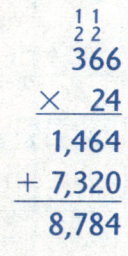

$$\begin{array}{r} \overset{\overset{1\,1}{2\,2}}{366} \\ \times\ \ 24 \\ \hline 1{,}464 \\ +\ 7{,}320 \\ \hline 8{,}784 \end{array}$$

Solution ____8,784 hours____

Why does the product of 366 and 2 tens have a 0 in the ones place?

Pair/Share
How are the regrouped ones and tens used to find the partial products?

---

7 A certain dishwasher uses 29 kilowatts of energy per hour. If the dishwasher was used 156 hours last year, how many kilowatts of energy did the dishwasher use last year?

Show your work.

What is the role of place value when multiplying two numbers?

Pair/Share
Show and explain how to solve this problem using a different method.

Solution _____

8 If 1 stamp costs 45 cents, what does a roll of 125 stamps cost in cents? What is this amount in dollars and cents?

Show your work.

> How could I first estimate the answer to this problem?

> **Pair/Share**
> Why is multiplication used to solve this problem? What other operation is used?

Solution _____

---

9 Raquel can type 63 words every minute. Rick can type 73 words every minute. How many more words can Rick type than Raquel in 135 minutes? Circle the letter of the correct answer.

A 1,350

B 4,599

C 8,505

D 9,855

Jared chose **B** as the correct answer. How did he get that answer?

_____

_____

_____

_____

> How can I use the difference in the number of words typed by Rick and Raquel every minute to solve this problem?

> **Pair/Share**
> Does Jared's answer make sense?

Lesson 5 Multiply Whole Numbers 43

Lesson 5   Independent Practice

Practice Multiplying Whole Numbers

Solve the problems.

1 Mrs. Cady constructs a cube with 512 magnetic blocks. Students in her two classes will each make an identical cube. There are 28 students in one class and 25 students in the other class. How many blocks does she need for all her students?

   A 5,120

   B 12,800

   C 14,336

   D 27,136

2 What are the values of the regrouped amounts in the multiplication below?

$$\begin{array}{r} \overset{2\,3}{435} \\ \times\ \ 17 \\ \hline 3{,}045 \\ +\ 4{,}350 \\ \hline 7{,}395 \end{array}$$

   A 2 and 3

   B 20 and 3

   C 200 and 30

   D 2,000 and 300

3 Choose *Yes* or *No* to tell whether the expression is equivalent to 179 × 44.

   a. 179 × (4 + 4)   ☐ Yes   ☐ No

   b. (179 × 40) + (179 × 4)   ☐ Yes   ☐ No

   c. (100 × 4) + (70 × 4) + (9 × 4)   ☐ Yes   ☐ No

   d. 4,000 + 2,800 + 360 + 400 + 280 + 36   ☐ Yes   ☐ No

   e. (100 × 44) + (70 × 44) + (9 × 44)   ☐ Yes   ☐ No

4 Show two different ways to complete the multiplication problem.

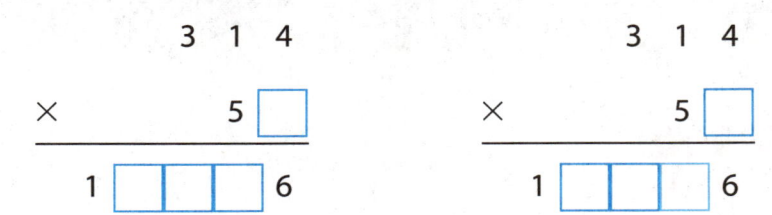

```
    3 1 4              3 1 4
  ×     5 ☐          ×     5 ☐
  ─────────          ─────────
  1 ☐ ☐ 6            1 ☐ ☐ 6
```

5 At the start of the day there are 78 boxes of DVDs in a warehouse. Each box has 116 DVDs. Then 19 of the boxes are shipped. Now how many DVDs are left in the warehouse?

Show your work.

Answer _____ DVDs

6 Use the distributive property two different ways to find the product of 127 and 32.

Show your work.

✓ **Self Check** Go back and see what you can check off on the Self Check on page 1.

Lesson 6 Introduction
Divide Whole Numbers

NC.5.NBT.6

Use What You Know

In the last lesson, you learned how to find products of two- and three-digit factors. Now you will learn how to divide with two-digit divisors.

There are 345 fifth graders enrolled at Wilson Middle School and 15 fifth-grade classrooms. How many students are in each class if each class has the same number of students?

a. What multiplication equation can you use to solve 345 ÷ 15? _____

b. Multiply 15 by multiples of 10. Fill in the blanks.

15 × 10 = _____   15 × 20 = _____   15 × 30 = _____

c. Now estimate the quotient. The quotient will be between which two tens? _____

d. If 15 × 20 = 300, what number is left after you subtract this product from 345? _____

e. Divide what is left by 15. _____ ÷ 15 = _____

f. Use the information above to find 345 ÷ 15. Explain your thinking.

_____

_____

_____

_____

46 Lesson 6 Divide Whole Numbers

Find Out More

On the previous page, you used the relationship between multiplication and division along with properties of operations to divide.

You can also use an area model to show division. It is similar to an area model for multiplication. You can think of division as finding a missing factor. The dividend is the product, the divisor is the known factor, and the quotient is the unknown factor.

The area model shows 345 ÷ 15.

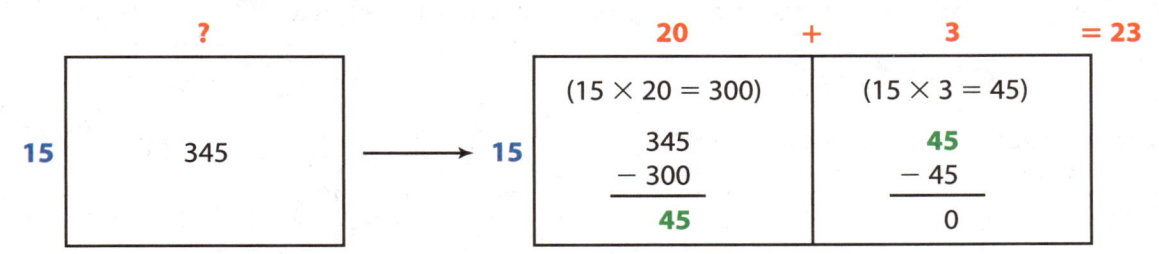

```
        ?                                    20        +       3       = 23
  ┌──────────┐                          ┌──────────┬──────────┐
  │          │                          │(15 × 20  │(15 × 3   │
15│   345    │             →         15 │  = 300)  │  = 45)   │
  │          │                          │   345    │    45    │
  └──────────┘                          │ − 300    │  − 45    │
                                        │ ─────    │  ────    │
                                        │    45    │     0    │
                                        └──────────┴──────────┘
```

Start with the greatest place and find the quotient place by place. On the previous page, you estimated the quotient to be between 20 and 30. So, **20** is the greatest ten that can be in the quotient.

- Start by multiplying **15** by **20**.
- Subtract the product from the dividend, 345.
- Write the difference, **45**, in the next section of the model.
- Think: **15 × ? = 45**. Multiply **15** by **3** and subtract this product from **45**.
- The difference is 0, so there is nothing left to divide.

▶ Reflect

1 How is dividing with an area model similar to multiplying with an area model? How is it different?

_____

_____

_____

_____

Lesson 6 Divide Whole Numbers

Lesson 6   Modeled and Guided Instruction

Learn About Dividing by Two-Digit Numbers

Read the problem below. Then explore different ways to divide by a two-digit divisor.

> A grocery store only sells eggs by the dozen. There are 12 eggs in 1 dozen eggs. If there are 624 eggs in stock, how many dozens of eggs are there?

▶ **Model It** You can use the relationship between multiplication and division to estimate the quotient in a division problem with a two-digit divisor.

$624 \div 12 = ?$ and $12 \times ? = 624$

Multiply 12 by multiples of 10. Make a table.

Number of dozens	10	20	30	40	50	60
Number of eggs	120	240	360	480	600	720

Since 624 is between **600** and **720**, the quotient is between **50** and **60**.

▶ **Model It** You can use an area model to solve a division problem with a two-digit divisor.

	?
12	624

→

50 + **2** = **52**

| 12 | $(12 \times 50 = 600)$
 624
 − 600
 24 | $(12 \times 2 = 24)$
 24
 − 24
 0 |

48   Lesson 6 Divide Whole Numbers

Connect It Now you will connect the area model to partial quotients.

2. How many hundreds are in the dividend? _____
How many groups of 12 are in 600? _____
This partial quotient is written above the bar.
What equation in the area model shows this?

_____

```
      52  ← quotient
       2  ← partial quotient
      50  ← partial quotient
12)624
   −600
     24
    −24
      0
```

3. Why is 600 subtracted from 624?
_____

4. How does the area model relate to finding the second partial quotient?
_____
_____

5. Explain how to use the partial quotients to find 624 ÷ 12. _____
_____

6. Describe how to divide using partial quotients. _____
_____
_____
_____
_____

Try It Use what you just learned about dividing by a two-digit divisor to solve these problems. Show your work on a separate sheet of paper.

7. An elementary school has 1,134 seeds. The seeds will be planted in 27 rows. Each row will have the same number of seeds. How many seeds will be planted in each row? _____

8. Nathan writes an essay with a total of 583 words. Every line of his essay has 11 words. How many lines long is Nathan's essay? _____

Lesson 6 Divide Whole Numbers 49

Lesson 6 Modeled and Guided Instruction

Learn About **Finding Quotients With Remainders**

Read the problem below. Then explore different ways to find quotients with remainders.

> Students are packing 140 oranges in boxes. They put 25 oranges in each box. How many boxes can they fill?

▶ **Model It** **You can use repeated subtraction to divide.**

$$
\begin{array}{r}
25\overline{)140} \\
-25 \quad \text{1 group of 25} \\
\hline
115 \\
-25 \quad \text{1 group of 25} \\
\hline
90 \\
-25 \quad \text{1 group of 25} \\
\hline
65 \\
-25 \quad \text{1 group of 25} \\
\hline
40 \\
-25 \quad \text{1 group of 25} \\
\hline
15 \leftarrow \text{remainder}
\end{array}
$$

▶ **Model It** **You can use a number line to divide.**

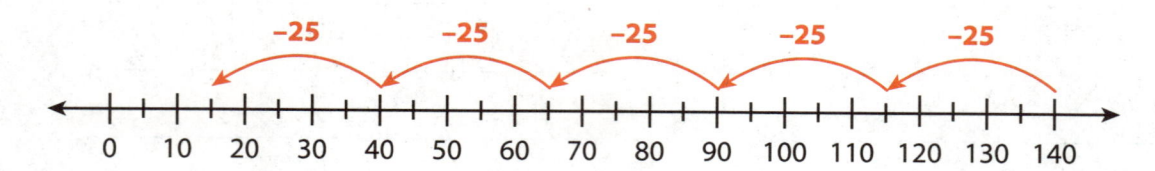

Connect It Now you will compare using partial quotients and using subtraction to divide.

9 Look at the first *Model It*. How many groups of 25 were subtracted from 140? Explain how you know. _____

_____

10 Look at the first *Model It*. Why is there a remainder?

_____

_____

11 Look at the second *Model It*. How many jumps of 25 are shown on the number line? _____ Where does the last jump on the number line end at? _____ What is the quotient of 140 ÷ 25? _____

12 Explain how the remainder relates to the problem of determining the number of boxes students can fill with oranges.

_____

_____

13 How many boxes can students completely fill with oranges? _____

14 Describe how to divide using repeated subtraction.

_____

_____

_____

Try It Use what you just learned about finding quotients with remainders to solve this problem. Show your work on a separate sheet of paper.

15 A sports team has $390 to spend on banners. Each banner costs $35. What is the greatest number of banners they can buy? How much money will they have left?

_____

Lesson 6 Divide Whole Numbers **51**

Lesson 6 Guided Practice

Practice: Dividing Whole Numbers

Study the example below. Then solve problems 16–18.

Example

Each package has 21 pieces of chalk. How many packages can be made with 1,704 pieces of chalk?

Look at how you could show your work using partial quotients.

```
        81 R 3
         1 ← partial quotient
        80 ← partial quotient
    21)1,704
      − 1,680 ← 21 × 80
          24
        − 21 ← 21 × 1
           3 ← remainder
```

Solution __81 packages__

Why is 80 a good number to use as a first partial quotient?

Pair/Share
How can you use multiplication to check that the quotient is correct?

---

16 A water cooler holds 1,284 ounces of water. How many more 6-ounce glasses than 12-ounce glasses can be filled from a full cooler?

Show your work.

You can first estimate how many glasses of each size can be filled.

Pair/Share
Explain how you found your estimate.

Solution _____

52 Lesson 6 Divide Whole Numbers

17 Students are planting 116 flowers in pots. Each pot can hold 18 flowers. How many pots do the students need to plant all of the flowers?

Show your work.

What numbers can I use to estimate the quotient?

Pair/Share
Explain how to check the answer to a division problem.

Solution _____

18 Harrison creates balloon animals for different events. He has 6,442 balloons. He wants to use the same number of balloons for each of 28 events. How many balloons can Harrison use at each event? Circle the letter of the correct answer.

A 203

B 230

C 231

D 232

What will be the greatest place in the quotient?

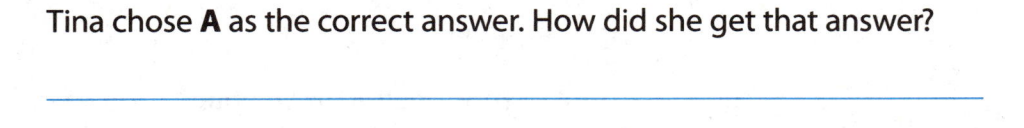

Tina chose **A** as the correct answer. How did she get that answer?

_____

_____

_____

Pair/Share
Does Tina's answer make sense?

©Curriculum Associates, LLC   Copying is not permitted.   Lesson 6 Divide Whole Numbers   **53**

Lesson 6 Independent Practice

Practice: Dividing Whole Numbers

Solve the problems.

1 Which equation can NOT be represented by the model below?

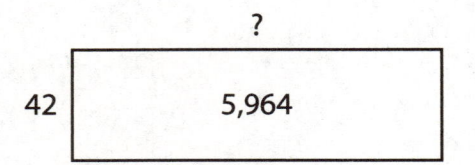

| 42 | 5,964 |

with ? above the bar

A 5,964 − ? = 42

B 5,964 ÷ ? = 42

C 42 × ? = 5,964

D 5,964 ÷ 42 = ?

2 Lisa's camera has 2,050 megabytes of memory for storing pictures. She has already used half this amount. A high-quality picture uses 16 megabytes of memory. How many high-quality pictures can Lisa store with the remaining memory?

_____

3 The tablets Mrs. King wants to buy for her class cost $42 each. She has $518. How many tablets can she buy?

Show your work.

Solution _____

4 Use the grid to draw a rectangle with an area of 1,125 square units and a side of 25 units.

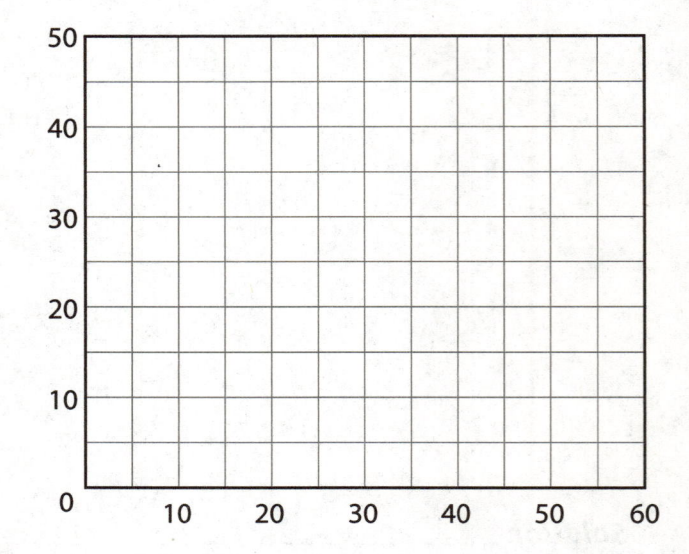

5 Vera makes a table to help solve the problem 673 ÷ 16. Which is the best estimate of the quotient?

| 10 | 20 | 30 | 40 | 50 | 60 |
| 160 | 320 | 480 | 640 | 800 | 960 |

A a number between 30 and 40

B a number close to 40

C about 52

D a number between 50 and 60

54 Lesson 6 Divide Whole Numbers

6 Mr. Kovich writes the problem 32 × △ = 1,696 on the board. Write a division equation that can be used to find the value of the triangle, and then find the value of the triangle.

Show your work.

Solution _____

7 Mr. Sullivan is organizing teams for the middle school's annual field day. There are eight classes at the school and 22 students in each class.

Part A What is the total number of students at the school?

Answer _____ students

Part B Mr. Sullivan wants to have 12 students on each team. How many full teams of 12 will there be? How many students will not be one full team?

Answer _____ teams _____ students

Part C How many students could Mr. Sullivan put on each team so that all students would be on a team? How many teams would there be? Explain your answer using diagrams, pictures, mathematical expressions, and/or words.

Answer _____ students _____ teams

✓ **Self Check** Go back and see what you can check off on the Self Check on page 1.

Lesson 7   Introduction
Add and Subtract Decimals

NC.5.NBT.7

Use What You Know

In grade 4, you learned to add whole numbers by adding the values of digits with the same place value. Now you'll add decimals the same way. Take a look at this problem.

> Sabrina and Christie are running in a relay. Sabrina runs 100 meters in 13.25 seconds, and Christie then runs the same distance in 12.2 seconds. What is their total time?

a. About how many seconds did the girls run altogether? _____ Explain your reasoning.
_____
_____

b. You can think of Sabrina's time as 1 ten + 3 ones + 2 tenths + 5 hundredths. Write Christie's time in the same way. _____

c. Combine Sabrina's and Christie's times. How many tens in all? _____ How many ones in all? _____ How many tenths in all? _____ How many hundredths in all? _____

d. Write the sum of the tens, ones, tenths, and hundredths as a decimal. _____

e. How does the sum compare to your estimate? Is your answer reasonable?
_____

f. Use words to explain how you could find Sabrina and Christie's total time.
_____
_____

56   Lesson 7 Add and Subtract Decimals   ©Curriculum Associates, LLC   Copying is not permitted.

▶▶ Find Out More

You can use what you know about adding whole numbers to add decimals. To add 13.25 and 12.2, you combine like place values.

One way to add decimals is to stack the numbers vertically. Lining up the decimal points is a way to keep track of place values. Using a place-value chart can help.

	Tens	Ones	.	Tenths	Hundredths
	1	3	.	2	5
+	1	2	.	2	0
	2	5	.	4	5

$\frac{2}{10}$ is equivalent to $\frac{20}{100}$, so you can write a 0 in the hundredths column.

The total time is 25.45 seconds.

To find the difference between Sabrina's and Christie's times, subtract 12.20 from 13.25. You can use what you know about subtracting whole numbers to subtract decimals.

	Tens	Ones	.	Tenths	Hundredths
	1	3	.	2	5
−	1	2	.	2	0
	0	1	.	0	5

Christie was faster by 1.05 seconds.

▶ Reflect

1 If Sabrina's time were 13.26 seconds instead of 13.25 seconds, would that change your estimate for the total time? Explain. _____

_____

_____

_____

_____

Lesson 7 Add and Subtract Decimals

Lesson 7 Modeled and Guided Instruction

Learn About: Adding Decimals to Thousandths

Read the problem below. Then explore different ways to understand how to add decimals to solve the problem.

> From his home, Tim rides the bus 3.982 miles. Then he walks 0.04 mile from the bus stop to school. How many miles does Tim travel from home to school?

▶ **Picture It** You can picture adding two decimals on a number line.

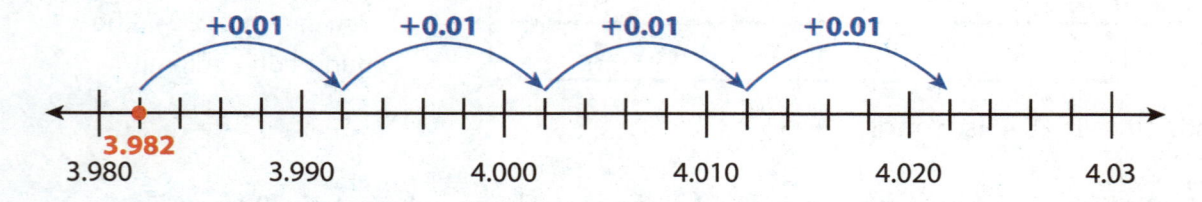

A number line from 3.980 to 4.03 shows four jumps of +0.01 starting at 3.982, landing at 4.022.

Starting at 3.982, you can make 4 jumps of 0.01 to the right to show the sum of 3.982 and 0.04.

▶ **Model It** You can use a place-value chart to help you understand how to add decimals.

	Ones	.	Tenths	Hundredths	Thousandths
Bus ride	3	.	9	8	2
Bus stop to school	0	.	0	4	0

The sum is 3 ones + 9 tenths + 12 hundredths + 2 thousandths.

$\frac{4}{100}$ is equivalent to $\frac{40}{1,000}$, so you can write a 0 in the thousandths column.

58 Lesson 7 Add and Subtract Decimals

► **Connect It** Now you will use the picture and the model to help you understand how to add decimals.

2 How can you use the number line in *Picture It* to figure out how many miles Tim travels from home to school? _____

_____

_____

3 Look at *Model It* on the previous page. What is another way to express 12 hundredths? _____

What is another way to express the sum? _____

4 You can add the decimals without a place-value chart by stacking them vertically. Line up the decimal points to keep track of place values.

$$\begin{array}{r} 3.982 \\ + 0.040 \\ \hline \end{array}$$

Why do you align the 8 in 3.982 with the 4 in 0.040?

_____

5 The addition problem to the right is partially completed. Explain why there is a 1 above the tenths place.

$$\begin{array}{r} \overset{1\ 1}{3.982} \\ + 0.040 \\ \hline .022 \end{array}$$

_____

_____

6 Complete the problem. How many miles does Tim travel from home to school?

_____

7 Explain how to add decimals.

_____

_____

► **Try It** Use what you just learned about adding decimals to solve this problem. Show your work on a separate sheet of paper.

8 Yana made a trail mix with 128.25 grams of dried fruit and 41.813 grams of almonds. How many grams of trail mix did Yana make? _____

Lesson 7 Add and Subtract Decimals **59**

Lesson 7 Modeled and Guided Instruction

Learn About Subtracting Decimals to Thousandths

Read the problem below. Then explore different ways to subtract decimals.

> Marty cuts 2.057 pounds of cheese from a 4.125-pound block of cheddar cheese. How many pounds of cheddar cheese are left in the block?

▶ **Picture It** You can subtract decimals using base-ten models.

Model 4 wholes, 1 tenth, 2 hundredths, and 5 thousandths.

▶ **Model It** You can subtract decimals using a place-value chart.

Using a place-value chart helps you make sure the place values are lined up correctly.

	Ones	.	Tenths	Hundredths	Thousandths
Original amount	4	.	1	2	5
Amount cut	2	.	0	5	7

60 Lesson 7 Add and Subtract Decimals ©Curriculum Associates, LLC Copying is not permitted.

Connect It Now you will use the base-ten models and the place-value chart to understand how to subtract decimals.

9 Look at the base-ten models on the previous page. Explain why parts of the models are crossed out. _____

_____

10 According to *Picture It*, how many pounds of cheddar cheese are left in the block?

_____

11 Look at the place-value chart in *Model It* on the previous page. Why can't you just subtract the 7 from the 5? _____

_____

_____

12 You can subtract the decimals without a place-value chart by stacking them vertically. Line up the decimal points to keep track of place values. You can express 2 hundredths as _____ + 10 thousandths.

$$\begin{array}{r} 4.125 \\ -\ 2.057 \end{array}$$

13 You can rewrite the problem as

$$\begin{array}{r} \overset{11}{} \\ \overset{0}{}\overset{\cancel{1}15}{} \\ 4.\cancel{1}\cancel{2}\cancel{5} \\ -\ 2.057 \end{array}$$

Complete the subtraction problem. There are _____ pounds of cheddar left.

14 Explain how to subtract two decimals and how to tell if your answer is reasonable.

_____

_____

_____

Try It Use what you just learned about subtracting decimals to solve this problem. Show your work on a separate sheet of paper.

15 Gwen is sending a box to a friend. The box weighs 23.54 pounds. Gwen removes a 4.476-pound book to decrease the shipping cost. What is the new weight of the box?

_____

Lesson 7 Modeled and Guided Instruction

Learn About: Estimating Decimal Sums and Differences

Read the problem below. Then explore different ways to estimate decimal sums and differences.

> Paula and Diego are building a clubhouse. They cut one piece of wood that is 2.471 meters long and another that is 2.428 meters long. Diego asked Paula about how long is the combined length of the two pieces of wood is. Estimate the sum to the nearest tenth to approximate the combined length.

▶ **Picture It** You can use rounding to estimate.

Place 2.471 and 2.428 on a number line to show their relationship to tenths.

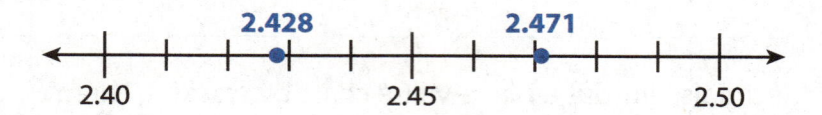

```
        2.428           2.471
◄───┼───┼───●───┼───┼───┼───●───┼───┼───┼───►
   2.40            2.45            2.50
```

2.471 is more than halfway between 2.400 and 2.500, so you can round it up to the nearest tenth, **2.5**.

2.428 is less than halfway between 2.400 and 2.500, so you can round it down to the nearest tenth, **2.4**.

You can add 2.5 and 2.4 to estimate the sum.

▶ **Model It** You can use front-end estimation with rounding.

Add the front end numbers, which are the whole numbers.	Round the decimals to the nearest tenth and add.
2.471 + 2.428 ――― 4	2.**471** → 0.5 + 2.**428** → 0.4 ―――――― 0.9

You can add 4 and 0.9 to estimate the sum.

62 Lesson 7 Add and Subtract Decimals

©Curriculum Associates, LLC Copying is not permitted.

▶ **Connect It** Now you will use the models on the previous page to estimate the sum.

16. What numbers do you add when you use rounding to estimate the sum of 2.471 and 2.428? _____

17. What is an estimate for the sum of 2.471 and 2.428 using rounding? _____

18. What numbers do you add when you use front-end estimation and rounding to estimate the sum of 2.471 and 2.428? _____

19. What is an estimate for the sum of 2.471 and 2.428 using front-end estimation and rounding? _____

20. What is the sum of 2.471 and 2.428? Are the estimates in problems 3 and 5 reasonable? Explain. _____
_____
_____

21. Explain how to estimate the sum or difference of decimals.
_____
_____

▶ **Try It** Use what you just learned about estimating sums and differences to solve these problems. Show your work on a separate sheet of paper.

22. Students are riding a bus to summer camp. The distance to camp is 68.749 kilometers. They rode 31.23 kilometers so far. Estimate to the nearest tenth how many kilometers they are from summer camp.

    _____ kilometers

23. Jay bought a fishing rod for $24.50 and a tackle box for $9.89. Estimate to the nearest dollar how much he spent.

_____

Lesson 7 Add and Subtract Decimals

Lesson 7 Guided Practice

Practice: Adding and Subtracting Decimals

Study the example below. Then solve problems 24–26.

Example

Diana has 3 different beads on her necklace. The red bead is 0.684 centimeter long, the multi-colored bead is 1.222 centimeters long, and the blue bead is 0.8 centimeter long. What is the total length of the beads on Diana's necklace?

Look at how you could show your work using equations.

$$\begin{array}{r} \overset{1}{1}.222 \\ + 0.684 \\ \hline 1.906 \end{array} \qquad \begin{array}{r} \overset{1}{1}.906 \\ + 0.800 \\ \hline 2.706 \end{array}$$

$$0.682 + 1.222 + 0.8 = 2.706$$

Solution ____2.706 centimeters____

The student needed two steps to solve the problem.

Pair/Share
Does it matter in what order you add the decimals?

---

24 On average, outdoor cats live 3.182 years and indoor cats live 16.71 years. How much longer does an average indoor cat live than an average outdoor cat?

Show your work.

How many hundredths are equivalent to 7 tenths?

Pair/Share
How do you know what operation to use to solve this problem?

Solution _____

Lesson 7 Add and Subtract Decimals

25 Kenton is shopping for clothes at a twelfth anniversary sale. He buys a pair of jeans priced at $24.99 and a clearance-priced shirt for $5.25. The store reduces the amount of his entire purchase by $12.12. How much does Kenton pay for his clothes? Show how to use estimation to tell if your answer is reasonable.

Show your work.

> This problem takes more than one step to solve.

Solution _____

> **Pair/Share**
> What step do you do first? What step do you do next?

26 Three boxes of cereal have masses of 379.41 grams, 424.256 grams, and 379.37 grams. What is the difference between the box of cereal with the greatest mass and the box of cereal with the least mass?

A 44.646 grams

B 44.846 grams

C 44.886 grams

D 45.286 grams

Cambria chose **D** as the correct answer. How did she get that answer?

_____

_____

_____

_____

> **What operation will solve this problem?**

> **Pair/Share**
> How could Cambria have checked her answer?

©Curriculum Associates, LLC   Copying is not permitted.

Lesson 7 Add and Subtract Decimals

Lesson 7 — Independent Practice

Practice: Adding and Subtracting Decimals

Solve the problems.

1 Randy rode his bike 1.238 miles to school from his house. After school, he rode 0.9 mile farther to the library. Randy biked home along the same route, stopping at a park 1.045 miles from the library. How many miles is the park from Randy's house?

- A 3.183
- B 1.945
- C 1.093
- D 0.688

2 Tim tracked the change in outside temperature one afternoon. He recorded a temperature of 85.4°F at noon. The temperature then rose 3.85°F over the next 4 hours. At 5:00 PM, Tim recorded a temperature of 89.25°F. How did the temperature change between 4:00 PM and 5:00 PM?

- A The temperature increased 0.8°F.
- B The temperature decreased 0.2°F.
- C The temperature increased 1°F.
- D There was no change in temperature.

3 The Lakefront Hiking Trail is 7.546 kilometers longer than the Hillside Hiking Trail. The Hillside Trail is 13.28 kilometers long. Estimate the length of the Lakefront Trail to the nearest tenth.

- A 20.0 kilometers
- B 20.7 kilometers
- C 20.8 kilometers
- D 21.8 kilometers

4 Tell whether each equation is *True* or *False*.

a. $198.526 - 42.813 = 155.713$
   ☐ True   ☐ False

b. $73.27 + 251.6 = 98.43$
   ☐ True   ☐ False

c. $37.045 + 56.201 = 93.66$
   ☐ True   ☐ False

d. $70.64 - (9.3 + 29.36) = 90.7$
   ☐ True   ☐ False

e. $38.2 - (11.111 + 23.76) = 3.329$
   ☐ True   ☐ False

5 The sum of three decimal numbers is 6. Exactly one of the numbers is less than 1. What could the numbers be?

Show your work.

Solution _____

6 Choose all the models or expressions that represent the difference 2.37 − 0.102.

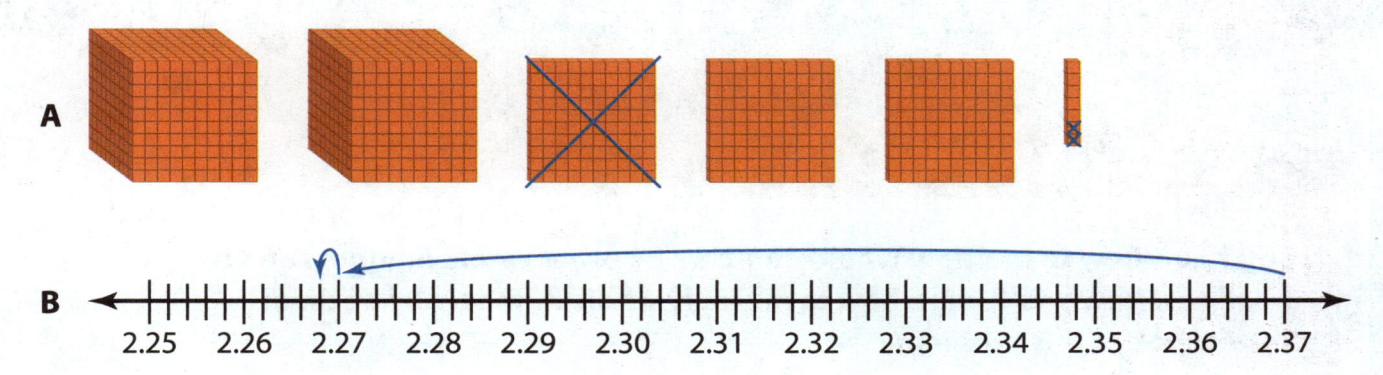

A

B

C 2 ones + 2 tenths + 6 hundredths + 8 thousandths

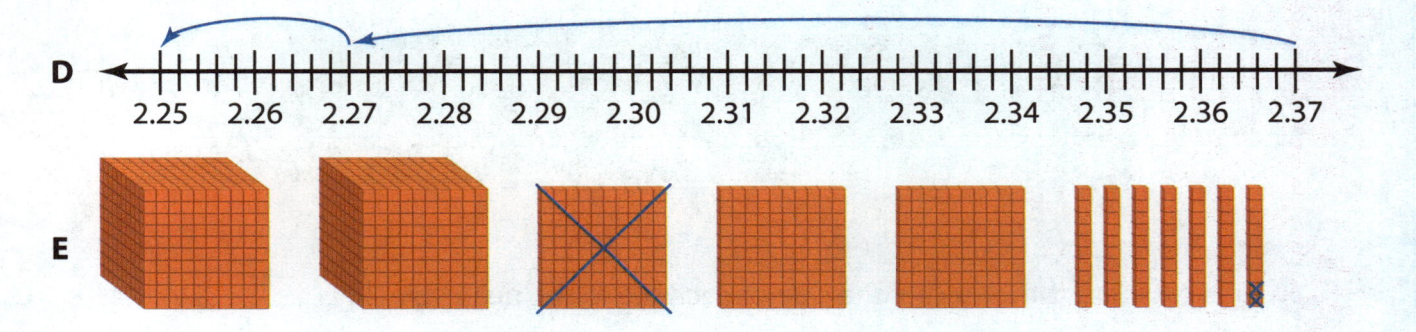

D

E

7 Ryan and Sarah are looking at cell phone plans. A group plan will cost $120.95 per month. An individual plan will cost $62.77 per month. Should Ryan and Sarah purchase a group plan or two individual plans? Justify your answer. How much money could they save?

Show your work.

✓ **Self Check** Go back and see what you can check off on the Self Check on page 1.

Lesson 8   Introduction
Multiply Decimals

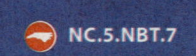

Use What You Know

You know how to multiply whole numbers by other whole numbers and by fractions. Now you'll learn how to multiply whole numbers by decimals. Take a look at this problem.

> Margo has 6 square tiles of equal size. Each side of each tile is 0.8 inch long. If Margo places all the tiles in a row with sides touching, how long is the row?

a. First, estimate the length of the side of each tile to the nearest inch. _____

b. Estimate the length of the row of 6 tiles using your answer. _____

c. Will the actual length be more or less than your estimate? _____

d. Why? _____

e. The length of the side of each tile is _____ tenths of an inch. There are _____ tiles in the row.

f. How many tenths of an inch long is the row of tiles? _____ tenths of an inch

g. Write the decimal equivalent:

The length of the row is _____ inches.

h. Is your answer reasonable? Explain your thinking.

_____

i. Use your own words to explain how you could find the length of the row of tiles.

_____

_____

68   Lesson 8 Multiply Decimals

Find Out More

On the previous page you saw that $6 \times 0.8 = 4.8$. How is this related to $6 \times 8 = 48$?

Notice that the digits of 6×0.8 are the same as the digits of 6×8. The digits of their products are also the same. Why is this?

In earlier lessons, you learned that dividing a number by 10 shifts the decimal point to the left so the value of the number decreases by a factor of 10. Dividing a number by 10 has the same result as multiplying the number by one tenth, or 0.1.

Look at the table below to see patterns when you multiply numbers by 0.1 and 0.01.

Expression	Equivalent Expressions	Product
6×8	$6 \times 8 \times 1.0$ 48×1.0	48.0
6×0.8	$6 \times 8 \times 0.1$ 48×0.1	4.8
6×0.08	$6 \times 8 \times 0.01$ $6 \times 8 \times 0.1 \times 0.1$ 48×0.01 $48 \times 0.1 \times 0.1$	0.48

Notice that the decimal point moves one place to the left each time you multiply by 0.1.

Reflect

1 What is the product of 6×0.008? Explain your reasoning.

_____

_____

_____

_____

_____

Lesson 8 Multiply Decimals

Lesson 8 Modeled and Guided Instruction

Learn About Estimating Decimal Products

Read the problem below. Then explore different ways to estimate decimal products.

> Tenesha walked for 2.25 hours. She walked 9.5 miles each hour. She multiplied 2.25 by 9.5 to find the total distance she walked. Her answer is 213.75 miles. Is 213.75 a reasonable product of 2.25 and 9.5? Use estimation to find out.

▶ Model It You can use rounding to estimate the product.

Round 2.25 and 9.5 to the nearest whole number.

2.25 → **2**

9.5 → **10**

You can multiply 2 by 10 to estimate the product.

▶ Model It You can use rounding to find numbers the product will fall between.

Round 2.25 and 9.5 down to the nearest whole number.

2.25 → **2**

9.5 → **9**

Round 2.25 and 9.5 up to the nearest whole number.

2.25 → **3**

9.5 → **10**

You can multiply 2 by 9 and 3 by 10 to find numbers the product will fall between.

Lesson 8 Multiply Decimals

▶ **Connect It** Now you will use the models on the previous page to estimate the product.

2 Look at the first *Model It*. What are 2.25 and 9.5 rounded to the nearest whole numbers? _____ What is the product of the rounded numbers? _____

3 Look at the second *Model It*. What are 2.25 and 9.5 rounded down to a whole number? _____ What are 2.25 and 9.5 rounded up to a whole number?

_____

4 Use the second *Model It*. Between what two numbers will the product fall?

_____

5 Use the estimates in problems 2 and 4 to explain if 213.75 is a reasonable product of 2.25 and 9.5. _____

_____

_____

6 Explain how to estimate the product of decimals. _____

_____

_____

_____

▶ **Try It** Use what you just learned about estimating decimal products to solve these problems. Show your work on a separate sheet of paper.

7 Students plan to buy 26 baseball caps for $9.49 each. Use estimation to find two products the total amount will be between. _____

8 A rectangular flower garden is 3.4 meters long and 12.7 meters wide. Estimate the area of the garden. _____

Lesson 8 Multiply Decimals

Lesson 8 Modeled and Guided Instruction

Learn About > Multiplying Decimals by Whole Numbers

Read the problem below. Then explore different ways to understand multiplying decimals by whole numbers

> Padma bought 3 pounds of grapes. Each pound of grapes costs $2.75. How much money did Padma spend on grapes?

▶ **Picture It** You can use an area model to picture multiplying decimals by whole numbers.

	2	0.7	0.05
3	3 × 2	3 × 0.7	3 × 0.05

▶ **Model It** You can use partial products to multiply decimals by whole numbers.

```
     2.75
  ×     3
   ───────
      15  ← 3 ones × 5 hundredths           =  15 hundredths
     210  ← 3 ones × 7 tenths = 21 tenths   = 210 hundredths
  + 600  ← 3 ones × 2 ones = 6 ones         = 600 hundredths
   ───────
     825 hundredths = 8.25
```

72 Lesson 8 Multiply Decimals

Connect It Use what you know about decimals and place value to solve the problem.

9 To solve the problem, you need to find 3 × $2.75. Estimate the total cost of the grapes. Explain your thinking. _____

_____

_____

10 Look at *Picture It*. How are the numbers above the area model related to 2.75? What do you multiply 2.75 by? _____

_____

_____

11 Look at *Model It*. How do the partial products relate to the area model?

_____

_____

12 What is the product of 3 and 2.75? _____ .

Is the product reasonable? Explain. _____

_____

13 How is multiplying with decimals like multiplying with whole numbers?

_____

_____

Try It Use what you have learned about multiplying decimals by whole numbers to solve these problems. Show your work on a separate sheet of paper.

14 Bananas cost $0.65 per pound. Sasha bought 4 pounds of bananas. How much did she pay for the bananas? _____

15 Brian tapes together 11 paper strips end to end. Each paper strip is 0.248 meter long. How long is the complete length of paper strips? _____

Lesson 8 Multiply Decimals   73

Lesson 8 Modeled and Guided Instruction

Learn About: Multiplying Decimals by Decimals

Read the problem below. Then explore different ways to multiply decimals

Hayden made a sign that is 1.4 meters long and 1.2 meters wide to post on the wall of his store. How many square meters of wall will the sign cover?

▶ **Picture It** You can use an area model to multiply decimals.

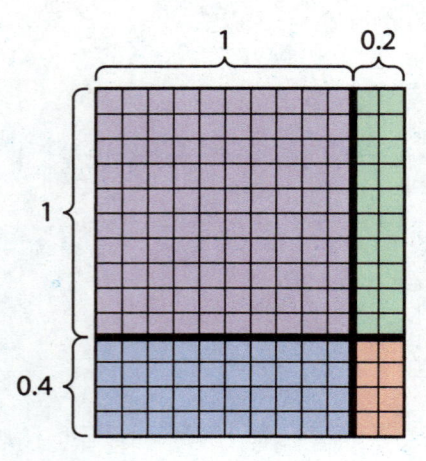

The rectangle measures **1.4 meters** by **1.2 meters**.

Each small square is 1 tenth of a meter by 1 tenth of a meter, or 0.1 meter by 0.1 meter.

The area of each small square is 0.1 meter × 0.1 meter = 0.01 square meter.

▶ **Model It** You can also use partial products to multiply decimals.

```
      1.2
    × 1.4
    -----
        8  ← 4 tenths × 2 tenths              =    8 hundredths
       40  ← 4 tenths × 1 one = 4 tenths      =   40 hundredths
       20  ← 1 one × 2 tenths = 2 tenths      =   20 hundredths
    + 100  ← 1 one × 1 one = 1 one            =  100 hundredths
    -----
      168 hundredths = 1.68
```

74 Lesson 8 Multiply Decimals

▶ **Connect It** Now you will use the model and properties of operations to solve the problem.

16 To solve the problem, you need to find 1.4 × 1.2. Estimate the area of wall that the sign will cover. Explain your thinking.

_____

17 Look at *Picture It*. Complete the area model below to find the area of each of the four sections of the rectangle.

1 × 1 = _____	1 × 0.2 = _____
0.4 × 1 = _____	0.4 × 0.2 = _____

18 Look at *Model It*. How do the partial products relate to the area model above?

_____

_____

19 Both *Picture It* and *Model It* show that 1.4 × 1.2 = _____ square meters. Is the product reasonable? Explain. _____

_____

_____

20 Explain what you know about the product when you multiply tenths by tenths.

_____

▶ **Try It** Use what you have learned about multiplying decimals to solve these problems. Show your work on a separate sheet of paper.

21 Rosa filled her car's tank with 9.8 gallons of gas. Each gallon costs $3.85. How much did Rosa spend on gas? _____

22 Harry has 0.5 bottle of water in his game bag. The bottle holds 0.95 liter of water. How many liters of water does Harry have? _____

Lesson 8 Multiply Decimals **75**

Lesson 8 · Guided Practice

Practice: Multiplying Decimals

Study the example below. Then solve problems 23–25.

Example

Liam ate 0.5 of a 1.25-ounce bag of raisins. How many ounces of raisins did Liam eat?

Look at how you could show your work using an area model.

	1	0.2	0.05
0.5	0.5 × 1 = 0.5	0.5 × 0.2 = 0.10	0.5 × 0.05 = 0.025

0.5 + 0.10 + 0.025 = 0.625

Solution ___0.625 ounce___

The student wrote 1.25 as 1 + 0.2 + 0.05 and used an area model to solve the problem.

Pair/Share
Solve the problem without a model.

---

23 Gina rides her bike to work at an average of 10.46 miles per hour. She bikes 1.2 hours each day. How many miles does Gina ride each day?

Show your work.

I multiply hundredths by tenths to solve this problem.

Pair/Share
What is a reasonable estimate for this problem? Explain your thinking.

Solution _____

76 Lesson 8 Multiply Decimals

24 If a person's hair grows 1.2 centimeters a month, how much would it grow in 9 months? Explain how to use an estimation strategy to check if your answer is reasonable.

Show your work.

Will the product be in tenths or hundredths?

*Pair/Share
Solve the problem using an area model.*

Solution _____

_____

_____

---

25 What is the product of 1.05 and 0.7? Circle the letter of the correct answer.

A 73.5

B 7.35

C 0.735

D 0.0735

Aaron chose **C** as the correct answer. How did he get that answer?

_____

_____

_____

_____

Will the product be greater than or less than 1.05?

*Pair/Share
Does Aaron's answer make sense?*

Lesson 8 Multiply Decimals

Lesson 8 **Independent Practice**

Practice: Multiplying Decimals

Solve the problems.

1 Terry finds the product of 4.27 and 5.1. Terry's answer is 21.777. Estimate the product of 4.27 × 5.1 to tell whether each sentence is *True* or *False*.

a. The product of 4.27 × 5.1 is between 9 and 11.   ☐ True   ☐ False

b. The product of 4.27 × 5.1 is between 20 and 30.   ☐ True   ☐ False

c. The product of 4.27 × 5.1 is about 200.   ☐ True   ☐ False

d. Terry's answer is reasonable.   ☐ True   ☐ False

2 Willa downloads 5 songs. Three of the song files are each 2.75 MB. Two song files are each 3.8 MB. How much space does Willa need for the songs she downloads?

A 5.55 MB

B 11.55 MB

C 15.85 MB

D 27.75 MB

3 Choose ALL the expressions that have the same value as the product of 0.11 and 4.5.

A 0.495 × 0.01

B 0.495 × 0.001

C 49.5 × 0.01

D 495 × 0.01

E 495 × 0.001

4 The area model below can be used to represent the product of 2.8 and 1.3. Complete the model by writing each of the following numbers in the correct part of the model.

0.6   2   0.24   0.8

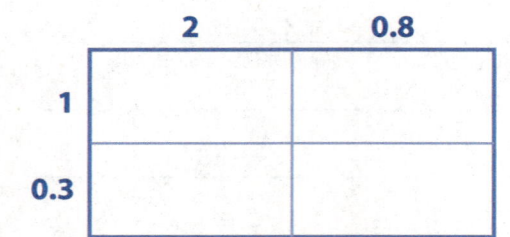

	2	0.8
1		
0.3		

78   Lesson 8 Multiply Decimals

5 Tyrone said that 2.35 × 5 equals 1.175 because there is only one digit before the decimal point in 2.35, so there must be one digit before the decimal point in the product. Use pictures, numbers, or words to explain whether or not Tyrone is correct.

Show your work.

_____

6 Each product below is missing a decimal point.

Part A Place the decimal point in each product so that the equation is correct.

12.53 × 5 = 6 2 6 5

4.28 × 3.6 = 1 5 4 0 8

1.3 × 0.89 = 1 1 5 7

7 × 6.12 = 4 2 8 4

Part B Circle one of the equations. Explain how you decided where to place the decimal point in this equation.

_____

_____

_____

_____

✓ **Self Check** Go back and see what you can check off on the Self Check on page 1.

Lesson 9 — Introduction
Divide Decimals

NC.5.NBT.7

Use What You Know

Now that you know how to multiply with decimals, you'll learn how to divide with decimals. Take a look at this problem.

> Mr. Kovich is preparing materials for a craft project. He needs to cut 2 meters of string into pieces that are 0.2 meter long. How many 0.2-meter pieces can he cut from 2 meters of string?

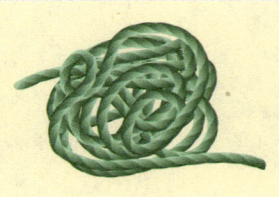

a. You need to find how many lengths of 0.2 meter are in 2 meters. Look at the drawing below.

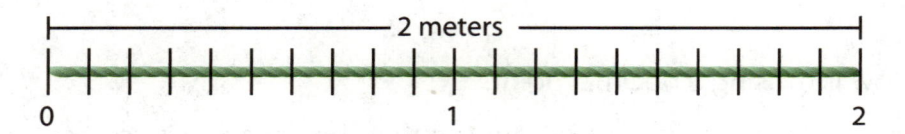

b. How many tenths of a meter are in 1 meter? _____

In 2 meters? _____

c. Circle lengths of 2-tenths meter on the drawing above.

How many lengths of 2-tenths meter are in 1 meter? _____

In 2 meters? _____

d. Explain how you could find how many 0.2-meter pieces can be cut from 2 meters of string.

_____

_____

_____

_____

80 Lesson 9 Divide Decimals   ©Curriculum Associates, LLC   Copying is not permitted.

Find Out More

On the previous page, you saw that 2 ÷ 0.2 = 10.
How is that similar to the division fact 2 ÷ 2 = 1?
How is it different?

Look at the table below to see some patterns in division.

Expression	Expression in words	Quotient
2 ÷ 2	How many groups of 2 are in 2?	1
2 ÷ 0.2	How many groups of 0.2 are in 2?	10
2 ÷ 0.02	How many groups of 0.02 are in 2?	100

When the divisor is greater than 1, the quotient is less than the dividend.

When the divisor is less than 1, the quotient is greater than the dividend.

dividend		divisor		quotient
2	÷	2	=	1
2	÷	0.2	=	10
2	÷	0.02	=	100

▶ Reflect

1 Look at the table above. What do you think is the quotient of 5 ÷ 0.05? Use a pattern to explain your reasoning.

_____

_____

_____

_____

_____

Lesson 9 Divide Decimals

Lesson 9 **Modeled and Guided Instruction**

Learn About Dividing a Whole Number by a Decimal

Read the problem below. Then explore different ways to divide a whole number by a decimal.

> Grant has 2 pounds of pretzels. He puts the pretzels into bags that each hold 0.25 pound. How many bags does Grant use to hold the pretzels?

▶ **Picture It** You can picture dividing a whole number by a decimal with decimal grids.

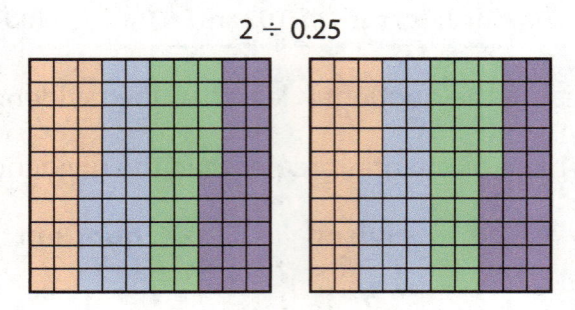

$2 \div 0.25$

▶ **Model It** You can use repeated subtraction to divide a whole number by a decimal.

```
0.25)2.00
     2.00
   − 0.25     1 group of 0.25
     1.75
   − 0.25     1 group of 0.25
     1.50
   − 0.25     1 group of 0.25
     1.25
   − 0.25     1 group of 0.25
     1.00
   − 0.25     1 group of 0.25
     0.75
   − 0.25     1 group of 0.25
     0.50
   − 0.25     1 group of 0.25
     0.25
   − 0.25     1 group of 0.25
        0
```

82   Lesson 9 Divide Decimals   ©Curriculum Associates, LLC   Copying is not permitted.

Connect It Now use decimal grids and repeated subtraction to solve the problem on the previous page.

2 $2 \div 0.25 = $ _____ ones ÷ _____ hundredths

3 Look at *Picture It*. How many hundredths are in the two grids? _____
How many hundredths are in each shaded group? _____
How many groups of hundredths are shaded? _____

4 Look at *Model It*. What number is divided into equal groups? _____
What number is repeatedly subtracted? _____
How many times is that number subtracted? _____

5 Explain how *Model It* is related to *Picture It*.

_____

_____

_____

_____

6 What is $2 \div 0.25$? How many bags does Grant use to hold the pretzels?

_____

7 How do you divide a whole number by a decimal? _____

_____

_____

_____

Try It Use what you just learned about dividing a whole number by a decimal to solve these problems. Show your work on a separate sheet of paper.

8 Students ran a total of 6 miles in a relay race. Each student ran 0.5 mile. How many students were in the race? _____

9 What is $3 \div 0.75$? _____

Lesson 9 Divide Decimals   83

Lesson 9 👥 Modeled and Guided Instruction

Learn About: Dividing a Decimal by a Whole Number

Read the problem below. Then explore different ways to divide a decimal by a whole number.

> Linda is making a denim jacket. She cuts 6.39 meters of fabric into 3 equal pieces. What is the length of each piece of denim?

▶ **Picture It** You can use an area model to picture dividing a decimal by a whole number.

	6 ÷ 3	0.3 ÷ 3	0.09 ÷ 3
3	6	0.3	0.09

▶ **Model It** You can use partial quotients to divide a decimal by a whole number.

```
   0.03     ← 9 hundredths ÷ 3 ones = 3 hundredths
   0.1      ← 3 tenths ÷ 3 ones = 1 tenth
   2        ← 6 ones ÷ 3 ones = 2 ones
 3)6.39
  −6
   0.3
  −0.3
   0.09
  −0.09
   0
```

84 Lesson 9 Divide Decimals   ©Curriculum Associates, LLC   Copying is not permitted.

Connect It Now you will use the area model and partial quotients to solve the problem on the previous page.

10 Look at *Picture It*. Why does the area model show three parts?

_____

_____

11 Look at *Picture It*. What do the expressions above the area model represent?

_____

_____

12 Look at *Model It*. What are the partial quotients of 6.39 ÷ 3? How are the partial quotients related to the expressions above the area model in *Picture It*?

_____

_____

13 What is the length of each piece of denim? _____

14 How is dividing a decimal by a whole number like dividing a whole number by a whole number? _____

_____

_____

Try It Use what you just learned about dividing a decimal by a whole number to solve these problems. Show your work on a separate sheet of paper.

15 Coach Ann is setting up a 5.6-kilometer race. She uses flags to mark off 8 equal sections of the race. How far apart should she space the flags to mark off the sections? _____

16 How much will each person receive if $20.35 is split evenly among 5 people?

_____

Lesson 9 Divide Decimals **85**

Lesson 9 · **Modeled and Guided Instruction**

Learn About ▸ Estimating Quotients

Read the problem below. Then explore different ways to estimate decimal quotients.

> Rachel bought 5 cans of paint. Each can is the same size. The total weight of the paint is 56.5 pounds. Rachel divided 56.5 by 5 to find the weight of each can of paint. Her answer is 11.3 pounds. Is 11.3 a reasonable quotient for 56.5 divided by 5? Use estimation to find out.

▶ **Model It** You can use rounding to estimate a quotient.

Round 56.5 to the nearest ten.

56.5 → 60

You can divide 60 by 5 to estimate the quotient.

▶ **Model It** You can use compatible numbers to estimate a quotient.

56.5 is close to 55 and 55 is easy to divide by 5.

You can divide 55 by 5 to estimate the quotient.

Connect It Now you will use the models on the previous page to solve the problem.

17 Look at the first *Model It*. Why does rounding 56.5 to the nearest ten make it easier to divide by 5? _____

_____

18 What is the estimate of 56.5 ÷ 5 when 56.5 is rounded to the nearest ten? _____

19 Look at the second *Model It*. Why is 55 used for the dividend to estimate the quotient?

_____

_____

20 What is the estimate of 56.5 ÷ 5 when 55 is used for the dividend? _____

21 Use the estimates in problems 18 and 20 to explain if 11.3 is a reasonable quotient of 56.5 ÷ 5. _____

_____

22 Explain how to estimate decimal quotients. _____

_____

_____

Try It Use what you just learned about estimating decimal quotients to solve these problems. Show your work on a separate sheet of paper.

23 The total cost for 6 pizzas is $41.94. Each pizza is the same price. Estimate the price of one pizza. _____

24 Mrs. Sanchez rode her bicycle 72 miles in 8.25 hours. If she rode the same number of miles each hour, about how many miles did she ride each hour?

_____

Lesson 9 Divide Decimals **87**

Lesson 9 — Guided Practice

Practice: Dividing Decimals

Study the model below. Then solve problems 26–28.

Example

Lionel uses a total of 2 meters of ribbon to wrap gifts. Each piece of ribbon is 0.4 meter long. How many gifts does Lionel wrap?

Look at how you could show your work using repeated subtraction.

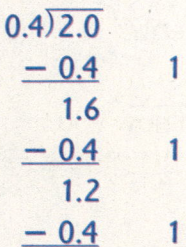

```
0.4)2.0
  − 0.4    1
    1.6
  − 0.4    1
    1.2
  − 0.4    1
    0.8
  − 0.4    1
    0.4
  − 0.4    1
      0    5
```

Solution Lionel wraps 5 gifts.

The student used repeated subtraction to divide 2 by 0.4.

Pair/Share
Can you solve the problem another way?

---

26 Nancy makes 7.84 ounces of soap for the school craft show. She pours all of the soap into 8 bottles. Each bottle has the same amount of soap. How many ounces of soap are in each bottle?

Show your work.

What number do you divide into equal groups? How many equal groups are there?

Pair/Share
How could you use an area model to solve the problem?

Solution _____

88 Lesson 9 Divide Decimals

27 Mrs. Jenkins has 6 laptops for her classroom. The total weight of the laptops is 28.9 pounds. Each laptop is the same weight. About how much does each laptop weigh?

Show your work.

Do you need an exact answer or an estimate?

Solution _____

Pair/Share
Explain how you solved the problem.

---

28 What is 10 ÷ 2.5? Circle the letter of the correct answer.

A 3

B 3.5

C 4

D 5

Gwen chose **A** as the correct answer. How did she get that answer?

_____
_____
_____

What methods could you use to answer the problem?

Pair/Share
How could you check the answer?

©Curriculum Associates, LLC   Copying is not permitted.

Lesson 9 Divide Decimals   **89**

Lesson 9 Independent Practice

Practice: Dividing Decimals

Solve the problems.

1 Jordan has $3 to spend at the used book store. Each comic book costs $0.60. How many comic books can Jordan buy?

   A 3

   B 4

   C 5

   D 6

2 Keith bought 3.4 pounds of peanuts on Monday, 2.5 pounds on Tuesday, and 4 pounds on Wednesday. He is going to divide the peanuts equally between himself and two friends. How many pounds of peanuts will each friend get?

   A 99 pounds

   B 33 pounds

   C 9.9 pounds

   D 3.3 pounds

3 A sticker is 1.2 inches wide. How many stickers will fit edge to edge on a strip of paper that is 6 inches wide?

   _____ stickers

4 Select *Yes* or *No* to tell if the quotient is about 4.

   a. $37.2 \div 9$ ☐ Yes   ☐ No

   b. $16 \div 4.2$ ☐ Yes   ☐ No

   c. $20.3 \div 7$ ☐ Yes   ☐ No

   d. $48 \div 8.8$ ☐ Yes   ☐ No

5 Jamie has 4 jars to fill with beads for a carnival game. She has 8.48 cups of multi-colored beads. Jamie wants to put an equal amount of beads in each jar. How many cups of beads can she place into each jar?

Part A Estimate the number of beads she can place into each jar.

Show your work.

Estimate _____ cups

Part B Complete the area model to solve the problem.

$8 \div 4 =$ _____ $0.4 \div 4 =$ _____ $0.08 \div 4 =$ _____

| 4 | 8 | 0.4 | 0.08 |

Answer _____ cups

Part C Use partial quotients to solve the problem. Explain if your estimate from Part A is reasonable.

Show you work.

Answer _____

_____

✓ **Self Check**   Go back and see what you can check off on the Self Check on page 1.

©Curriculum Associates, LLC   Copying is not permitted.   Lesson 9 Divide Decimals   **91**

Unit 1 MATH IN ACTION

Introduction
Use Whole Numbers and Decimals

SMP1 Make sense of problems and persevere in solving them.

Study an Example Problem and Solution

In this lesson, you will work with whole numbers and decimals to solve real-world problems. Look at this problem and one solution.

Dog Collars

Alex is organizing a pet fair. Money from the fair will be donated to the local pet shelter. Alex's friend Bella will have a booth at the fair. Here is Bella's sign.

Adorable Dog Collars

Small $8.99
Medium . . . $9.99
Large $10.99

Supplies to make a small collar cost $1.50. Bella estimates that the supplies for each size collar cost $0.50 more than the next smaller size. Bella hopes to make at least $200 for the pet shelter by selling collars.

- Find how much Bella makes on each collar after paying for supplies.
- Show a way to make at least $200 by selling collars.
- Include at least 5 collars of each size in your plan.

Read the sample solution on the next page. Then look at the checklist below. Find and mark parts of the solution that match the checklist.

Problem-Solving Checklist

☐ Tell what is known.
☐ Tell what the problem is asking.
☐ Show all your work.
☐ Show that the solution works.

a. **Circle** something that is known.
b. **Underline** something that you need to find.
c. **Draw a box around** what you do to solve the problem.
d. **Put a checkmark** next to the part that shows the solution works.

Alex's Solution

Hi, I'm Alex. Here's how I solved this problem.

▷ **I know the selling price** of each collar. **I need to find the cost** to make each size collar and subtract it from the price.

Size	Price	Cost of Supplies	Amount left to Donate
Small	$8.99	$1.50	$7.49
Medium	$9.99	$1.50 + $0.50	$7.99
Large	$10.99	$2.00 + $0.50	$8.49

I used a table to organize my information.

▷ **I'll round the donation amounts so that I can estimate.**
I can multiply each rounded amount by 5 since I have to include at least 5 of each size collar.

$7.49 ⟶ $7   and $7 × 5 = $35
$7.99 ⟶ $8   and $8 × 5 = $40
$8.49 ⟶ $8   and $8 × 5 = $40
                              $115

By rounding and estimating first, I avoid some computing with decimals.

▷ **I can see** that I need almost $90 more to get to $200. I can try to make about $30 with each of the three collar sizes.

$7 × 4 = $28 $8 × 4 = $32 $8 × 4 = $32

▷ **Now I can find the actual amounts from 5 + 4 small collars, 5 + 4 medium collars, and 5 + 4 large collars.**

$7.49 × 9 = (7 × 9) + (0.4 × 9) + (0.09 × 9) and 63 + 3.6 + 0.81 = 67.41
$7.99 × 9 = (7 × 9) + (0.9 × 9) + (0.09 × 9) and 63 + 8.1 + 0.81 = 71.91
$8.49 × 9 = (8 × 9) + (0.4 × 9) + (0.09 × 9) and 72 + 3.6 + 0.81 = 76.41
                                                                215.73

If Bella sells 9 of each size collar she will be able to donate $215.73.

The solution works because it includes at least 5 collars of each size and it makes more than $200.

This is my final answer.

Unit 1 Math in Action   Modeled and Guided Instruction

Try Another Approach

There are many ways to solve problems. Think about how you might solve the Dog Collars problem in a different way.

Dog Collars

Alex is organizing a pet fair. Money from the fair will be donated to the local pet shelter. Alex's friend Bella will have a booth at the fair. Here is Bella's sign.

Adorable Dog Collars

Small $8.99
Medium . . . $9.99
Large $10.99

Supplies to make a small collar cost $1.50. Bella estimates that the supplies for each size collar cost $0.50 more than the next smaller size. Bella hopes to make at least $200 for the pet shelter by selling collars.

- Find how much Bella makes on each collar after paying for supplies.
- Show a way to make at least $200 by selling collars.
- Include at least 5 collars of each size in your plan.

▶ **Plan It** Answer these questions to help you start thinking about a plan.

A. How can you use estimation to help find a solution?

_____

_____

B. How can you use the sample solution to plan how to find a different solution?

_____

_____

▶ **Solve It** Find a different solution for the Dog Collars problem. Show all your work on a separate sheet of paper.

You may want to use the problem-solving tips to get started.

Problem-Solving Tips

- **Models** You may want to use . . .
  - partial products.
  - area models.

- **Word Bank**

round	multiply	greater than
estimate	product	decimal

- **Sentence Starters**
  - I can round _____
  - If I multiply _____

Problem-Solving Checklist
Make sure that you . . .
- ☐ tell what you know.
- ☐ tell what you need to do.
- ☐ show all your work.
- ☐ show that the solution works.

▶ **Reflect**

Use Mathematical Practices As you work through the problem, discuss these questions with a partner.

- **Use Models** How can you use an area model to help find a solution?
- **Repeated Reasoning** If you are multiplying by 0.99, how can you multiply by a whole number and use subtraction to find the product?

Unit 1 Math in Action — Guided Practice

Discuss: Models and Strategies

Read the problem. Write a solution on a separate sheet of paper. Remember, there can be lots of ways to solve a problem!

Petting Zoo

The zoo where Alex works has agreed to bring some animals to the pet fair. Guests can pay to feed and play with the animals. Alex has to decide which animals to bring and how much food they will need.

Petting Zoo Notes
- Include 2 or 3 different kinds of animals.
- Include more than 10 but fewer than 20 animals.
- Have enough food to feed each animal a day's worth of food.

Alex reads the keeper's notes to find about how much an average animal eats in a day.

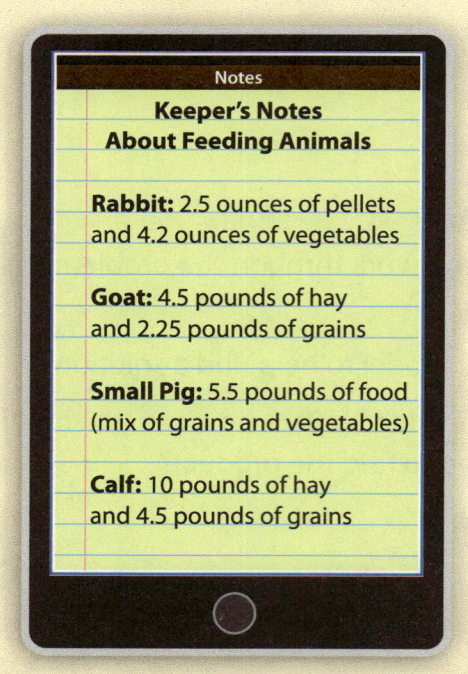

Keeper's Notes About Feeding Animals

Rabbit: 2.5 ounces of pellets and 4.2 ounces of vegetables

Goat: 4.5 pounds of hay and 2.25 pounds of grains

Small Pig: 5.5 pounds of food (mix of grains and vegetables)

Calf: 10 pounds of hay and 4.5 pounds of grains

How much food should Alex bring to the pet fair?

▶ **Plan It and Solve It** Find a solution to the Petting Zoo problem.

Write a detailed plan for Alex. Be sure to include:
- which animals and how many of each to bring to the pet fair.
- a list of the food she will need to feed all the animals for 1 day, including amounts of each kind of food.
- reasons for the choices you made.

Problem-Solving Tips

- **Questions**
  - How does the size of the animal affect your choice?
  - Which animals eat some of the same kind of food?

- **Sentence Starters**
  - I would bring _____
  - Alex needs about _____

Problem-Solving Checklist

Make sure that you . . .
☐ tell what you know.
☐ tell what you need to do.
☐ show all your work.
☐ show that the solution works.

▶ **Reflect**

Use Mathematical Practices As you work through the problem, discuss these questions with a partner.

- **Make Sense of Problems** What will you do first? Why?
- **Make an Argument** How can you justify the choices you made?

Unit 1 Math in Action Independent Practice

Persevere ▸ On Your Own

Read the problems. Write a solution on a separate sheet of paper. Remember, there are many different ways to solve a problem!

Robot Area

Alex asked her friend Beau to bring his robots to the fair. Guests can buy tickets to play with the robots. Beau needs to rope off an area of the fairgrounds to keep his robots in sight. Read his notes.

Robot Area Notes
- The area should be rectangular.
- It needs to be more than 100 feet long and less than 100 feet wide.
- The area needs to be between 7,500 and 10,000 square feet.

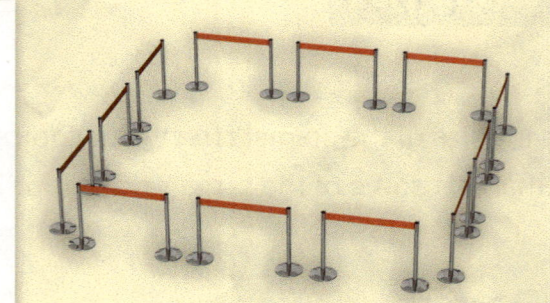

What dimensions should the robot area have?

▶ **Solve It** Describe an area that Beau can rope off for his robots.
- Give the length and width of the area.
- Give the area in square feet.
- Explain why your measurements work.

▶ **Reflect**

Use Mathematical Practices After you complete the task, choose one of these questions to discuss with a partner.

- **Use Structure** How did you use place value ideas to think about numbers that would work?
- **Persevere** Did you try different combinations of numbers before deciding on a final answer? Explain.

Barely Used

Alex's friend Brandi has a collection of used books, CDs, and DVDs that people have given her. She will sell them at a booth at the pet fair. Brandi thinks she can sell enough items to make at least a $100 donation to the pet shelter. Look at Brandi's prices.

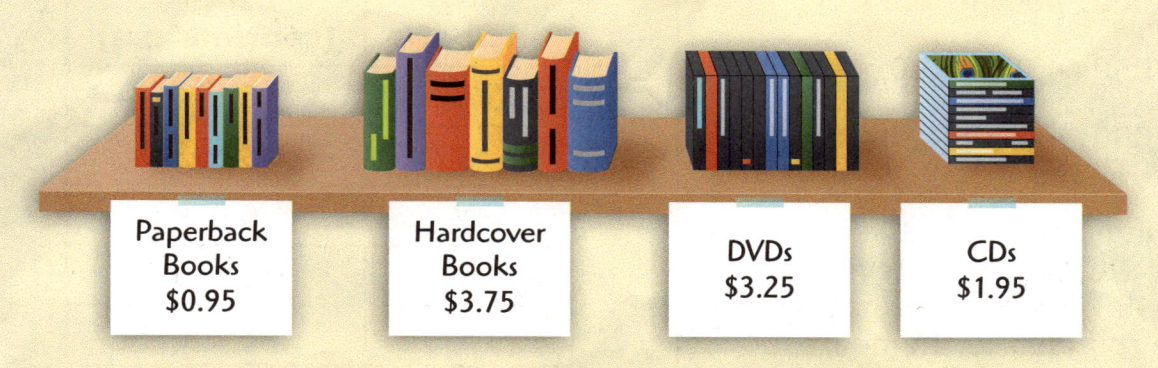

Paperback Books	Hardcover Books	DVDs	CDs
$0.95	$3.75	$3.25	$1.95

What items can Brandi sell to make at least $100?

▶ **Solve It** Find a combination of items that Brandi needs to sell to meet her goal.
- Include some of all four items.
- For each item, tell how many Brandi needs to sell and how much money she will make.
- Give the total amount Brandi will make selling all the items on your list.
- Explain why this total works.

▶ **Reflect**

Use Mathematical Practices After you complete the task, choose one of these questions to discuss with a partner.
- **Use Models** What equations or expressions did you use to find your solution?
- **Persevere** What were all of the steps you took to find a solution?

Unit 1 Assessment
Interim Assessment

Solve the problems.

1. Which pattern shows multiplying by 0.01?

   A 40,000; 400; 40; 4

   B 40,000; 400; 4; 0.04

   C 40,000; 4,000, 400, 4

   D 40,000; 4,000; 4; 0.04

2. The driving distance between Durham, NC, and Fayetteville, NC, is 143.456 kilometers. What is this distance rounded to the nearest tenth?

   A 143 kilometers

   B 143.5 kilometers

   C 143.46 kilometers

   D 140 kilometers

3. Julie likes to ride her bike during summer vacation. In June, she rode her bike 104.789 miles. In July, she rode 129.532 miles. In August, she rode 61.055 miles. Which statement is true?

   A Julie rode more than twice as many miles in July as in August.

   B Julie rode a total of 295.326 miles during summer vacation.

   C Julie rode 43.744 miles more in June than in August.

   D Julie rode more miles in June than in July.

4. The table shows the lengths of the five keys that Paco has on his keychain.

Key Label	Length (centimeters)
T	6.104
G	5.793
W	6.079
H	5.785
S	6.106

   Which inequality is true?

   A T > S

   B W < G

   C G > H

   D S < W

5. What is the value of 381 × 27?

100   Unit 1 Interim Assessment   ©Curriculum Associates, LLC   Copying is not permitted.

6   Ben writes the number 44,444. The value of the digit 4 in the hundreds place is 10 times the value of the digit 4 in what place value?

   A   ones place

   B   tens place

   C   thousands place

   D   ten thousands place

7   Ben's teacher asks him to write a three-digit number that uses the digit 3 once and the digit 6 twice. The value of one 6 needs to be $\frac{1}{10}$ the value of the other 6. What number can Ben write?

8   Coach Miller is having a cookout for his baseball team. He bought 5.15 pounds of ground beef to make burgers. The cost of the ground beef is $3.40 for each pound. He used all the ground beef to grill 17 burgers. Which equation can he use to find the cost of each burger?

   A $3.40 \times 5.15 - 17 = 0.51$

   B $5.15 \times 3.40 \div 17 = 1.03$

   C $5.15 - 3.40 \div 17 = 4.95$

   D $3.40 \div 17 + 5.15 = 5.35$

Unit 1 Interim Assessment continued

Performance Task

Answer the questions and show all your work on separate paper.

You have a movie theater gift card worth $40, so you invite a friend to go to the movies with you. Your friend challenges you to spend the exact value of the gift card. Find at least one way to do so by choosing from the items listed below.

Item	Price
2D movie ticket	$7.25
3D movie ticket	$8.50
Small Popcorn	$3.25
Medium Popcorn	$5.50
Large Popcorn	$7.75
Bottle of Water	$2.50
1 Ounce of Snack Mix	$0.50

Checklist
Did you . . .
- [ ] organize your choices?
- [ ] check your calculations?
- [ ] use words and numbers to complete the task?

▶ **Reflect**

Use Mathematical Practices After you complete the task, choose one of the following questions to answer.

- **Persevere** What strategies did you use? If a strategy did not work, what did you do?
- **Model** How did you use equations to solve this problem?

Unit 2
Number and Operations—Fractions

Let's learn about multiplying and dividing fractions.

Real-World Connection Some situations require sharing. Suppose you and a friend make a batch of 24 cookies to share equally with your families. Dividing the batch into two groups allows you each to have 12 cookies. But you might decide that since your friend has a large family, your friend should get $\frac{2}{3}$ of the batch. How many cookies does your friend get? If your friend gets $\frac{2}{3}$ of the batch of cookies, can you also get $\frac{2}{3}$ of the batch? In order to decide, you need to know how to multiply by a fraction and how to add fractions.

In This Unit You will learn how to add, subtract, multiply, and divide fractions. You will also see how fractions are used in many real-world situations.

✓ Self Check

Before starting this unit, check off the skills you know below. As you complete each lesson, see how many more skills you can check off!

I can:	Before this unit	After this unit
add and subtract fractions with unlike denominators, for example: $\frac{5}{8} + \frac{1}{4} = \frac{7}{8}$.	☐	☐
estimate sums or differences of fractions, for example: $2\frac{3}{8} + 5\frac{1}{2}$ is a little less than 8.	☐	☐
multiply fractions, for example: $\frac{2}{3} \times \frac{3}{4} = \frac{6}{12}$, or $\frac{1}{2}$.	☐	☐
divide with unit fractions, for example: $4 \div \frac{1}{7} = 28$.	☐	☐

Lesson 10 Introduction
Add and Subtract Fractions

NC.5.NF.1

Use What You Know

In Grade 4, you learned that adding and subtracting fractions is similar to adding and subtracting whole numbers. Take a look at this problem.

Emiliano needs $\frac{1}{2}$ stick of butter to make corn bread. He also needs $\frac{1}{4}$ stick of butter to make apple muffins. How many sticks of butter does he need?

$\frac{1}{2}$ stick of butter

$\frac{1}{4}$ stick of butter

BUTTER   BUTTER

a. Does Emiliano need more than 1 stick of butter or less than 1 stick of butter for the corn bread and the apple muffins? _____

b. How do you know? _____
_____

c. Do $\frac{1}{2}$ and $\frac{1}{4}$ have like denominators? _____

d. How do you find equivalent fractions? _____
_____

e. 4 is a multiple of both 2 and 4. Write an equivalent fraction for $\frac{1}{2}$ using fourths.

$\frac{1}{2} = \frac{\Box}{4}$

f. You now have fractions that have same-size parts. How many fourths of a stick of butter does Emiliano need?

$\frac{1}{2} + \frac{1}{4} =$ _____ $+ \frac{1}{4} =$ _____ . So he needs _____ stick of butter.

104 Lesson 10 Add and Subtract Fractions   ©Curriculum Associates, LLC   Copying is not permitted.

▶▶ Find Out More

To add fractions, the size of the parts that make up the whole must be the same. Because $\frac{1}{2}$ and $\frac{1}{4}$ use different-size parts, you rewrote $\frac{1}{2}$ as the equivalent fraction $\frac{2}{4}$. Since $\frac{1}{4}$ and $\frac{2}{4}$ both use fourths, you combined $\frac{2}{4}$ and $\frac{1}{4}$ to get the sum $\frac{3}{4}$.

The same idea works for subtracting fractions. Here is a similar problem: $\frac{3}{4} - \frac{1}{2}$. Write $\frac{1}{2}$ as the equivalent fraction $\frac{2}{4}$.

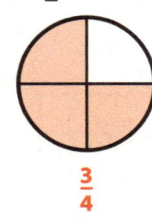

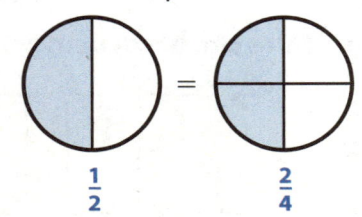

$\frac{3}{4}$ $\frac{1}{2}$ = $\frac{2}{4}$

To subtract $\frac{2}{4}$ from $\frac{3}{4}$, you can take 2 shaded parts away from the 3 shaded parts. You have 1 shaded part left.

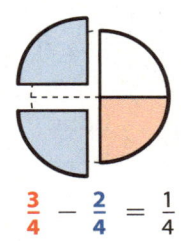

$\frac{3}{4} - \frac{2}{4} = \frac{1}{4}$

Before adding or subtracting fractions, the fractions must be built from the same-size part of a whole. This means the fractions must have the same denominator, or a **common denominator**.

▶ Reflect

1 Write a multiplication equation that shows how the denominators of $\frac{3}{4}$ and $\frac{1}{2}$ are related. Explain how this equation helps you write $\frac{1}{2}$ as an equivalent fraction with a denominator of 4.

_____

_____

_____

Lesson 10 Modeled and Guided Instruction

Learn About: Adding Fractions with Unlike Denominators

Read the problem below. Then explore different ways to add fractions with unlike denominators.

> Jenna spent $1\frac{2}{3}$ hours mowing the back yard. After taking a break, she spent $\frac{9}{12}$ hour mowing the front yard. How much time did she spend mowing the whole yard?

▶ **Picture It** You can picture the fractions in the problem using models.

The shaded parts represent time spent on the back yard, $1\frac{2}{3}$ hours, and the front yard, $\frac{9}{12}$ hour.

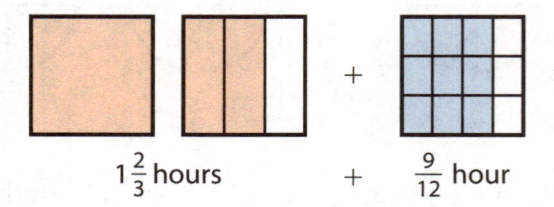

$1\frac{2}{3}$ hours   + $\frac{9}{12}$ hour

The sections need to be divided into same-size parts to add. Use dashed lines to divide the fraction model for the first fraction into 12 equal parts. This works because 12 is a multiple of 3.

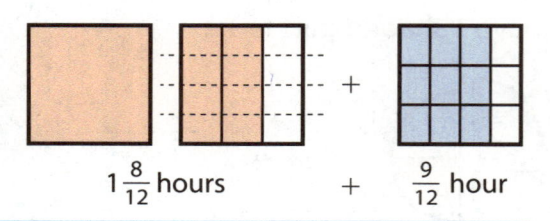

$1\frac{8}{12}$ hours   + $\frac{9}{12}$ hour

▶ **Model It** You can use a number line to estimate and add fractions.

The number line is divided into twelfths with points at $\frac{9}{12}$ and $1\frac{2}{3}$.

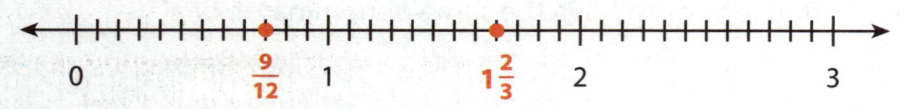

Estimate: The fraction $\frac{9}{12}$ is close to 1, and $1\frac{2}{3}$ is close to $1\frac{1}{2}$. So, the sum is about $2\frac{1}{2}$.

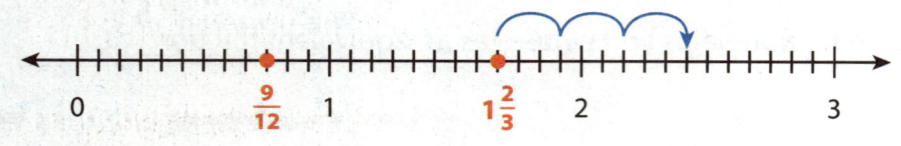

Add: Start at $1\frac{2}{3}$ and jump right $\frac{9}{12}$.

106 Lesson 10 Add and Subtract Fractions

Connect It Now you will solve the problem from the previous page using equivalent fractions and addition.

2 Are $1\frac{2}{3}$ and $\frac{9}{12}$ made up of same-size parts? Justify your answer.

_____

_____

_____

3 Look at the models on the previous page. What is a common denominator of $1\frac{2}{3}$ and $\frac{9}{12}$? _____

4 You can find this common denominator without a model. Write a multiplication equation that shows how the denominator 3 is related to 12. _____

5 Use this common denominator to find an equivalent fraction for $1\frac{2}{3}$.

$1\frac{2}{3} = 1$ _____

6 Using the equivalent fractions, what is the sum of $1\frac{2}{3}$ and $\frac{9}{12}$? _____

7 Is your answer reasonable? Explain. Use the estimate on the previous page.

_____

8 Explain how to add two fractions with unlike denominators. _____

_____

_____

_____

Try It Use what you just learned about adding fractions with unlike denominators to solve these problems. Show your work on a separate sheet of paper.

9 What is $1\frac{1}{4} + 2\frac{3}{8}$? _____

10 Hank practices $\frac{2}{5}$ of the words on his spelling list on Monday. He practices another $\frac{3}{10}$ of his list on Tuesday. How much of his spelling list has Hank practiced so far? _____

Lesson 10 Add and Subtract Fractions **107**

Lesson 10 Modeled and Guided Instruction

Learn About: Subtracting Fractions with Unlike Denominators

Read the problem below. Then explore different ways to subtract fractions with unlike denominators.

Gavin's water bottle has $1\frac{3}{8}$ cups of water. He drinks $\frac{1}{2}$ cup. How much water is left in the bottle?

▶ Picture It You can use a picture to model subtracting fractions.

The water bottle is shaded to show that it has 1 cup + $\frac{3}{8}$ cup of water in it. 1 cup is equivalent to $\frac{8}{8}$ cup. The bottle has 11 eighths shaded.

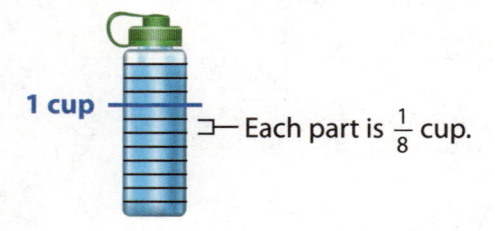

1 cup — Each part is $\frac{1}{8}$ cup.

Gavin drinks $\frac{1}{2}$ **cup**, or $\frac{4}{8}$ **cup**. So take away 4 shaded parts of the bottle to show how much is left.

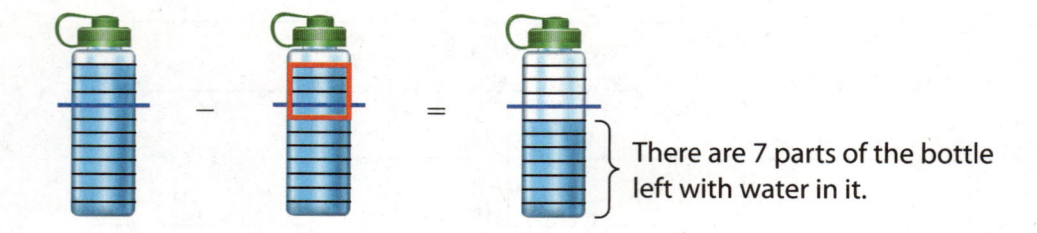

There are 7 parts of the bottle left with water in it.

▶ Model It You can use a number line to model subtracting fractions.

The number line below is divided into $\frac{1}{8}$s with a point at $1\frac{3}{8}$.

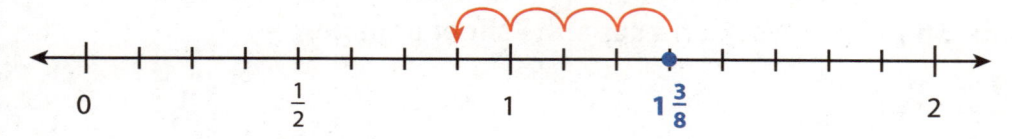

$\frac{1}{2}$ is equivalent to $\frac{4}{8}$. $\frac{4}{8}$ is four $\frac{1}{8}$ units on the number line. Start at $1\frac{3}{8}$ and jump left $\frac{4}{8}$.

108 Lesson 10 Add and Subtract Fractions

Connect It Now you will solve the problem from the previous page using equivalent fractions and a subtraction equation.

11 Estimate the amount of water left in Gavin's bottle. _____

How did you get your estimate? _____

_____

12 Look at *Picture It* and *Model It* on the previous page. Why is $\frac{1}{2}$ rewritten as $\frac{4}{8}$?

_____

_____

13 Use the number line on the previous page to rewrite $1\frac{3}{8}$ as a fraction greater than 1. Then use it to write a subtraction equation with equivalent fractions.

_____ − _____ = _____

14 Compare your answer to your estimate. Is your answer reasonable? How do you know?

_____

15 Why was it helpful to rewrite $1\frac{3}{8}$ as a fraction greater than 1? _____

_____

16 Explain how to subtract two fractions with unlike denominators. _____

_____

_____

Try It Use what you just learned about subtracting fractions with unlike denominators to solve these problems. Show your work on a separate sheet of paper.

17 What is $\frac{7}{8} - \frac{1}{2}$? _____

18 Emily's living room window is $2\frac{5}{6}$ feet wide. The window in her bedroom is $1\frac{1}{3}$ feet wide. How much wider is the living room window than her bedroom window? _____

Lesson 10 Add and Subtract Fractions

Lesson 10 — Guided Practice

Practice: Using Fractions with Unlike Denominators

Study the example below. Then solve problems 19–21.

Example

Chapter 1 of Henry's book is $5\frac{1}{3}$ pages long. Chapter 2 is $8\frac{1}{6}$ pages long. How much longer is Chapter 2 than Chapter 1?

Look at how you could show your work using an equation.

$$5\frac{1}{3} = 5 + \left(\frac{1 \times 2}{3 \times 2}\right) = 5\frac{2}{6}$$

$$8\frac{1}{6} = \left(7 + \frac{6}{6}\right) + \frac{1}{6} = 7\frac{7}{6}$$

$$7\frac{7}{6}$$
$$-5\frac{2}{6}$$
$$\overline{2\frac{5}{6}}$$

Solution $2\frac{5}{6}$ pages longer

I need to find a common denominator before I subtract fractions.

Pair/Share
Explain why the model regrouped $8\frac{1}{6}$ as $7\frac{7}{6}$.

---

19 What is $1\frac{3}{4} + 1\frac{1}{8}$? Estimate and then find the sum. Use your estimate to tell whether your answer is reasonable.

Show your work.

Use values that are close to the actual values to find an estimate.

Pair/Share
Explain why you chose the type of model you did to solve this problem.

Solution _____

110 Lesson 10 Add and Subtract Fractions

20 Michael rode his bike $2\frac{2}{3}$ miles on Saturday. He rode another $1\frac{5}{6}$ miles on Sunday. How many miles did Michael ride his bike on both days combined?

Show your work.

How are the denominators 3 and 6 related?

Solution _____

Pair/Share
How did you decide what operation to use to solve this problem?

---

21 Cara's bathroom floor has an area of $2\frac{3}{8}$ square yards. She lays down a rug that has an area of $1\frac{1}{4}$ square yards. What area of floor is NOT covered by the rug? Circle the letter of the correct answer.

A 1 square yard

B $1\frac{1}{8}$ square yards

C $1\frac{2}{8}$ square yards

D $1\frac{4}{12}$ square yards

John chose **D** as the correct answer. How did he get that answer?

_____

_____

_____

What equivalent mixed numbers could I subtract?

Pair/Share
Does a denominator of 12 make sense?

Lesson 10 Add and Subtract Fractions **111**

Lesson 10 Independent Practice

Practice: Using Fractions with Unlike Denominators

Solve the problems.

1. The model below represents the expression $1\frac{5}{8} + 2\frac{1}{2}$.

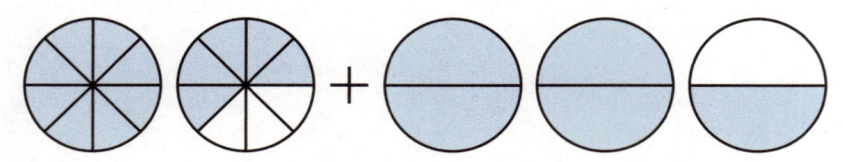

   Which of the following could NOT be represented by the model?

   A $1 + 2 + \frac{6}{8}$

   B $1\frac{5}{8} + 2\frac{4}{8}$

   C $1 + 2 + \frac{9}{8}$

   D $\frac{13}{8} + \frac{20}{8}$

2. Mackenzie's footprint is $\frac{7}{12}$ foot long. Her dad's footprint is $1\frac{1}{6}$ feet long. Which equation can be used to find how much longer Mackenzie's dad's footprint is than Mackenzie's? Select *Yes* or *No* for each equation.

   a. $1\frac{2}{12} - \frac{7}{12} = ?$ ☐ Yes ☐ No

   b. $\frac{7}{12} + 1\frac{1}{6} = ?$ ☐ Yes ☐ No

   c. $1\frac{1}{6} - \frac{2}{3} = ?$ ☐ Yes ☐ No

   d. $\frac{14}{12} - \frac{7}{12} = ?$ ☐ Yes ☐ No

3. Find two fractions in the list below that can be added using the denominator 100. Write those two fractions in the box.

   $\frac{1}{5}$ $\frac{5}{6}$ $\frac{10}{12}$ $\frac{1}{8}$ $\frac{4}{10}$

Like denominator = 100

112 Lesson 10 Add and Subtract Fractions

4 Lucy is making a smoothie by following the recipe below.

Recipe
Sunshine Smoothie

$1\frac{1}{2}$ cups banana

$\frac{1}{2}$ cup yogurt

1 cup strawberries

$\frac{3}{4}$ cup orange juice

Place ingredients in blender.

Blend until smooth.

Part A Explain whether the recipe will make enough for Lucy and 3 friends to each have at least 1 cup of smoothie. If it doesn't make enough, explain how to change the recipe to make enough.

_____

_____

_____

_____

_____

Part B If Lucy uses the original recipe, how much smoothie remains after she gives each of her friends $\frac{3}{4}$ cup of smoothie?

Show your work.

Answer _____ cups

✓ **Self Check** Go back and see what you can check off on the Self Check on page 103.

©Curriculum Associates, LLC   Copying is not permitted.

Lesson 10 Add and Subtract Fractions

Lesson 11 Introduction
Add and Subtract Fractions in Word Problems

NC.5.NF.1

Use What You Know

Now that you can add and subtract fractions with different denominators, you can use this skill to solve word problems. Take a look at this problem.

Aleena has a 1-gallon watering can that is full of water. She uses $\frac{3}{8}$ gallon to water her roses and $\frac{1}{4}$ gallon to water her geraniums. How much water did Aleena use to water both the roses and geraniums?

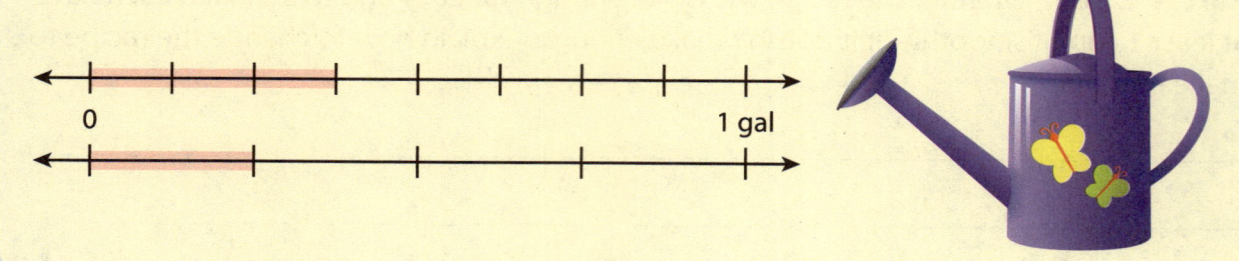

a. Does Aleena use more than $\frac{1}{2}$ gallon or less than $\frac{1}{2}$ gallon of water? _____

b. How do you know? _____

c. Estimate how much water Aleena used. _____

d. Write an equation with equivalent fractions to find the amount of water Aleena used.
_____

e. Is this answer reasonable based on your estimate? Explain. _____

114 Lesson 11 Add and Subtract Fractions in Word Problems

Find Out More

The number line below shows the location of **benchmark fractions** between 0 and 2. You can use these fractions to estimate sums and differences.

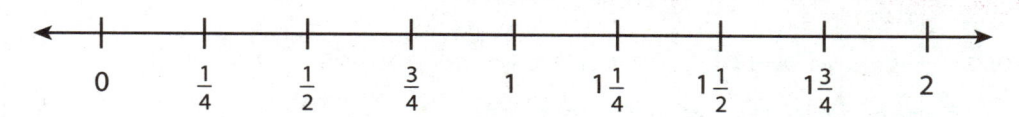

0 — $\frac{1}{4}$ — $\frac{1}{2}$ — $\frac{3}{4}$ — 1 — $1\frac{1}{4}$ — $1\frac{1}{2}$ — $1\frac{3}{4}$ — 2

There are different ways to estimate the sums or differences of fractions. The examples below show two ways to think about fractions to find estimates for the amount of water Aleena used to water her flowers.

$\frac{3}{8} + \frac{1}{4} = ?$

Student A

$\frac{3}{8}$ and $\frac{1}{4}$ are both less than $\frac{1}{2}$, or $\frac{4}{8}$.

So the sum must be less than 1.

Student B

$\frac{3}{8}$ is a little greater than $\frac{1}{4}$.
I know $\frac{1}{4} + \frac{1}{4}$ is $\frac{2}{4}$, or $\frac{1}{2}$.

So, I estimate the sum is greater than $\frac{1}{2}$.

Reflect

1 Both estimates above model correct thinking. Which estimate makes more sense to you? Why?

_____
_____
_____
_____

Lesson 11 Modeled and Guided Instruction

Learn About Solving Word Problems with Fractions

Read the problem below. Then explore different ways to estimate and solve problems with fractions.

> Frankie purchases a $3\frac{3}{6}$-pound bag of chicken. He uses $1\frac{1}{3}$ pounds of chicken for fajitas. How many pounds of chicken are left?

▶ **Picture It** You can picture the problem using a fraction strip.

The fraction strip below represents $3\frac{3}{6}$ pounds of chicken. It is separated into sections representing the $1\frac{1}{3}$ pounds used for fajitas and the unused amount.

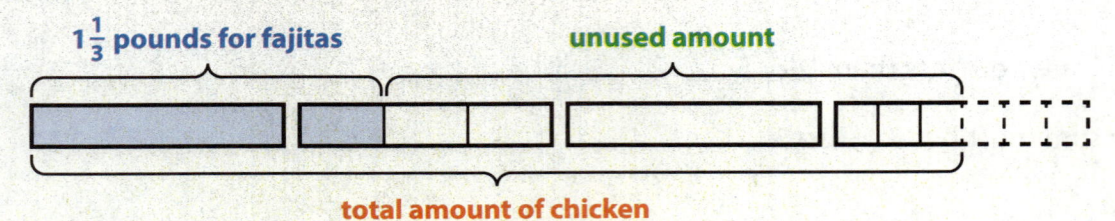

The model shows the difference $3\frac{3}{6} - 1\frac{1}{3}$.

▶ **Model It** You can model the problem with a number line.

Since 6 is a multiple of 3, the fractions in the problem, $3\frac{3}{6}$ and $1\frac{1}{3}$, can be rewritten using a common denominator of 6. $3\frac{3}{6} = 3\frac{3}{6}$, and $1\frac{1}{3} = 1\frac{2}{6}$.

The number line below is divided into sixths. It shows starting with a total of $3\frac{3}{6}$ pounds, with two jumps to the left representing the $1\frac{1}{3}$ pounds of chicken used.

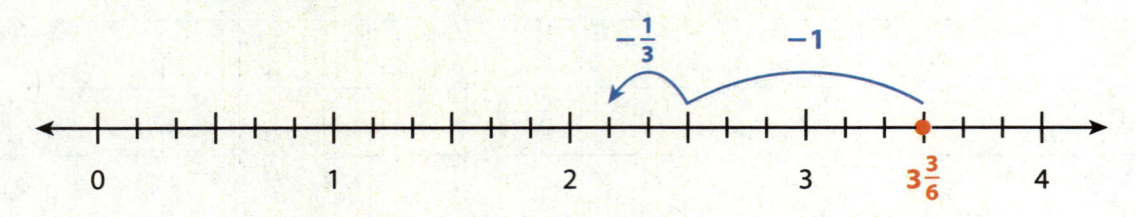

You can rewrite the difference $3\frac{3}{6} - 1\frac{1}{3}$ as $3\frac{3}{6} - 1\frac{2}{6}$.

116 Lesson 11 Add and Subtract Fractions in Word Problems

▶ **Connect It** Now you will estimate and then solve the problem from the previous page using benchmark fractions and an equation.

2 Identify the closest half on either side of $1\frac{1}{3}$.

$1\frac{1}{3}$ is greater than _____ and less than _____.

3 Why are halves a good choice for benchmark fractions for $1\frac{1}{3}$?

_____

4 The difference $3\frac{3}{6}$ minus $1\frac{1}{3}$ must be between _____ and _____.

Estimate $3\frac{3}{6} - 1\frac{1}{3}$ and explain your estimate.

_____

_____

5 Find the actual difference. _____

There are _____ pounds of chicken remaining.

6 Is this a reasonable answer based on your estimate? Explain. _____

_____

7 Explain how you can check if a fraction sum or difference is reasonable.

_____

_____

▶ **Try It** Use what you just learned about estimating with benchmark fractions to solve this problem. Show your work on a separate sheet of paper.

8 Tim's bean sprout grew $3\frac{3}{8}$ inches. Teegan's bean sprout grew $2\frac{3}{4}$ inches. How many more inches did Tim's bean sprout grow than Teegan's? First, estimate the difference and explain your reasoning. Then find the actual difference.

_____

_____

Lesson 11 Add and Subtract Fractions in Word Problems

Lesson 11 Modeled and Guided Instruction

Learn About Solving Two-Step Problems with Fractions

Read the problem below. Then explore different ways to solve two-step problems with fractions.

> Yasmin has $2\frac{1}{8}$ yards of fabric. She uses $\frac{7}{8}$ yard to make a tablecloth and $\frac{3}{4}$ yard to make napkins. How many yards of the fabric are left?

▶ **Picture It** You can picture the problem using fraction strips.

The fraction strips below represent $2\frac{1}{8}$ yards of fabric. They are divided into sections to represent the $\frac{7}{8}$ yard used for the tablecloth and the $\frac{3}{4}$ yard used to make napkins.

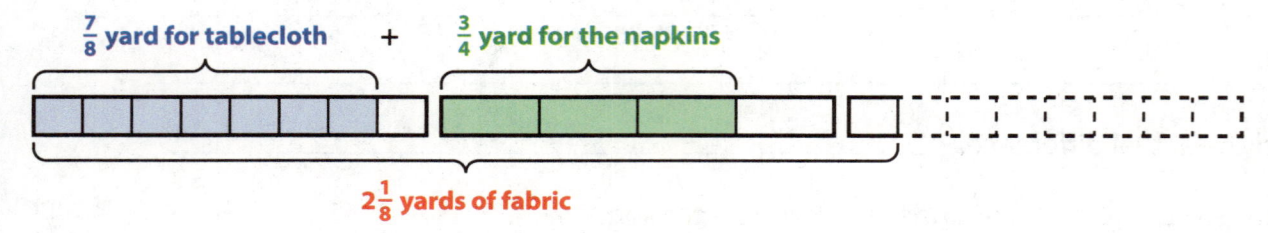

The picture shows the difference of $2\frac{1}{8}$ and the sum of $\frac{7}{8} + \frac{3}{4}$.

▶ **Model It** You can model the problem with a number line.

Since 8 is a multiple of 4, you can rewrite the fraction $\frac{3}{4}$ using a common denominator of 8. $\frac{3}{4} = \frac{6}{8}$

The number line below is divided into eighths. It shows starting with a total of $2\frac{1}{8}$ yards, with one jump to the left representing the $\frac{7}{8}$ yard used and another jump to the left representing the $\frac{3}{4}$ yard used.

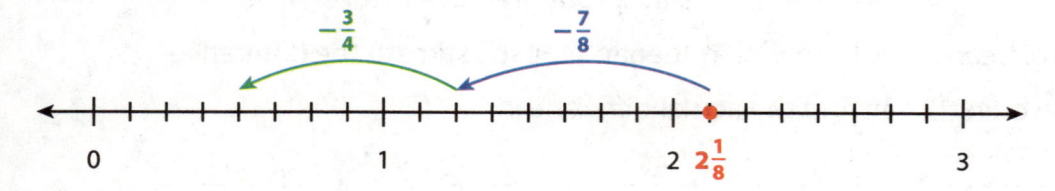

You can rewrite the problem $2\frac{1}{8} - \frac{7}{8} - \frac{3}{4}$ as $2\frac{1}{8} - \frac{7}{8} - \frac{6}{8}$.

118 Lesson 11 Add and Subtract Fractions in Word Problems

Connect It Now you will estimate and solve the problem from the previous page using number sense and an equation.

9 What operation or operations can you use to solve the problem?

_____

_____

10 Is there another way to solve the problem? Explain.

_____

_____

11 Use benchmark fractions to estimate the solution. Explain your estimate.

_____

_____

12 Find the sum $\frac{7}{8} + \frac{3}{4}$. _____

13 Use the number line on the previous page to rewrite the mixed number $2\frac{1}{8}$ as a fraction greater than 1. Then use it to write a subtraction equation to find the difference between the total yards of fabric and the yards of fabric used.

_____ − _____ = _____. There is _____ yard of fabric left.

14 Is this a reasonable answer based on your estimate? Explain.

_____

Try It Use what you just learned about solving two-step problems with fractions to solve this problem.

15 A punch bowl holds $3\frac{4}{12}$ quarts. Andy adds $1\frac{1}{3}$ quarts of orange juice and $\frac{5}{6}$ quart of seltzer water to the bowl. How many quarts of cranberry juice can Andy add to the bowl without the bowl overflowing? Estimate the amount and explain your reasoning. Then find the actual amount.

_____

_____

Lesson 11 Guided Practice

Practice: Solving Word Problems with Fractions

Study the example below. Then solve problems 16–18.

Example

The blue field of stars on a flag has an area of $1\frac{3}{5}$ square yards. The red stripes have a combined area of $2\frac{3}{10}$ square yards. What is the difference between the area of the blue field of stars and the area of the red stripes?

Look at how you could show your work using fraction strips.

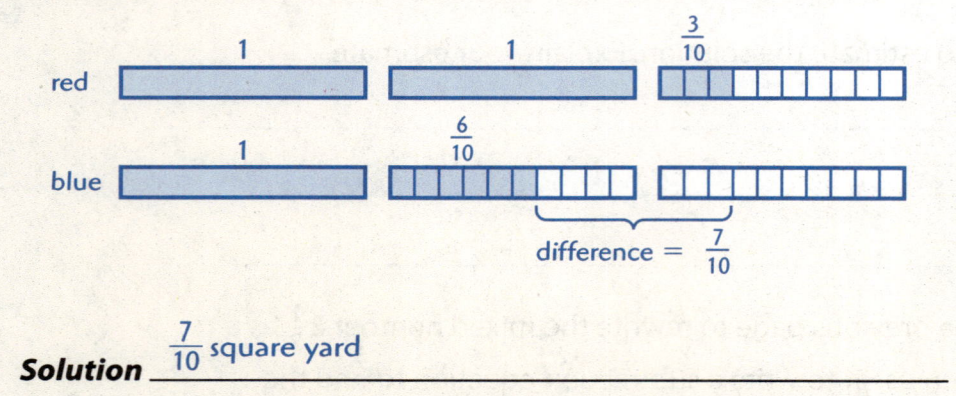

red: 1, 1, $\frac{3}{10}$

blue: 1, $\frac{6}{10}$

difference = $\frac{7}{10}$

Solution $\frac{7}{10}$ square yard

What common denominator was used to subtract these fractions?

Pair/Share
What benchmark fractions could you use to estimate the answer?

---

16 Parker mixes $3\frac{1}{2}$ ounces of blue paint with $1\frac{2}{8}$ ounces of yellow paint to create green paint to use for the leaves of a tree. How many ounces of green paint did Parker create?

Estimate, and then compute. Explain how you know your result is reasonable.

Show your work.

Will there be a little more than $4\frac{1}{2}$ ounces or a little less than $4\frac{1}{2}$ ounces of green paint?

Pair/Share
Was your estimate more than or less than the actual answer? By how much?

Solution _____

120 Lesson 11 Add and Subtract Fractions in Word Problems

17 Jose's football weighs $\frac{7}{10}$ pound. His football helmet weighs $3\frac{1}{10}$ pounds, and his shoulder pads weigh $3\frac{15}{100}$ pounds. Estimate how much more the helmet and ball weigh than the shoulder pads. Explain your estimate.

What benchmark fractions could I use to estimate the difference in weight?

Pair/Share
How does an estimate help you tell if your answer is reasonable?

Solution _____

18 Which is a reasonable estimate for the difference $5\frac{1}{2} - 3\frac{5}{8}$? Circle the letter of the correct answer.

A between $\frac{1}{2}$ and 1

B between 1 and $1\frac{1}{2}$

C between $1\frac{1}{2}$ and 2

D between 2 and $2\frac{1}{2}$

Elise chose **D** as the correct answer. How did she get that answer?

How can I use benchmark fractions to estimate the difference?

Pair/Share
Does Elise's answer make sense?

Lesson 11 Add and Subtract Fractions in Word Problems 121

Lesson 11 Independent Practice

Practice: Solving Word Problems with Fractions

Solve the problems.

1 William compares monthly rainfall amounts for the summer months using the table below.

Month	Monthly Rainfall
June	$3\frac{3}{8}$ inches
July	$3\frac{3}{4}$ inches
August	$3\frac{1}{2}$ inches

About how much more rain fell in June and July than in August?

A $3\frac{1}{4}$ inches

B $3\frac{1}{2}$ inches

C 4 inches

D $4\frac{1}{2}$ inches

2 Several expressions are shown. Decide if the value of each expression is less than $1\frac{1}{2}$, between $1\frac{1}{2}$ and 2, or greater than 2. Write each expression in the correct category in the chart.

$$2\frac{1}{2} - 1\frac{1}{8} \qquad 1\frac{5}{8} + \frac{3}{4} \qquad 3\frac{4}{5} - 1\frac{1}{10} \qquad \frac{2}{5} + \frac{9}{10}$$

Less than $1\frac{1}{2}$	Between $1\frac{1}{2}$ and 2	Greater than 2

122 Lesson 11 Add and Subtract Fractions in Word Problems

3 The table below shows the thickness of coins.

Coin	Thickness
quarter	$1\frac{75}{100}$ millimeters
dime	$1\frac{35}{100}$ millimeters
nickel	$1\frac{95}{100}$ millimeters
penny	$1\frac{5}{10}$ millimeters

Hailey stacks a dime on top of a penny. She estimates the thickness of the two coins to be less than 3 millimeters.

Write a symbol (<, >, or =) in the box to make the statement true. Then use the statement to tell whether Hailey's estimate is correct.

$1\frac{5}{10} + 1\frac{35}{100}$ ☐ $1\frac{5}{10} + 1\frac{5}{10}$

Is Hailey's estimate correct? _____

4 Jimmy says $3\frac{7}{12} - 2\frac{5}{6}$ is $1\frac{1}{3}$.

Part A Without finding the actual difference, explain why Jimmy's difference is or is not reasonable.

_____

_____

Part B Find the actual difference.

Show your work.

Solution _____

✓ **Self Check** Go back and see what you can check off on the Self Check on page 103.

Lesson 11 Add and Subtract Fractions in Word Problems **123**

Lesson 12 — Introduction
Fractions as Division

NC.5.NF.3

Use What You Know

You know that division is used for equal sharing and that fractions represent a number of equal parts. In this lesson, you will learn how division and fractions are related. Take a look at this problem.

Mrs. Tatum needs to share 4 fluid ounces of red paint equally among 5 art students. How much red paint will each student get?

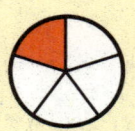

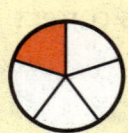

1 ounce   1 ounce   1 ounce   1 ounce

PAINT
4 ounces

a. Start with 1 ounce of paint. What fraction is represented by 1 ounce of paint shared equally among 5 students?

   1 ounce of paint shared equally among 5 students = _____ ounce per student

b. Look at the model above. Write a multiplication expression that represents a student getting $\frac{1}{5}$ of each of 4 ounces of paint.

   _____

c. What fraction is represented by the product? _____

d. 4 ounces of paint ÷ 5 students = _____ ounce of paint per student

▶▶ Find Out More

What if Mrs. Tatum wants to share 8 fluid ounces of paint equally among the 5 students? How much will each student get? You can think about this quotient two ways.

One way: Each student will get $\frac{1}{5}$ ounce of paint if 1 ounce of paint is shared among 5 students. There are 8 ounces to share.

1 ounce ÷ 5 students = $\frac{1}{5}$ ounce for each student

1 ounce 1 ounce 1 ounce 1 ounce 1 ounce 1 ounce 1 ounce 1 ounce

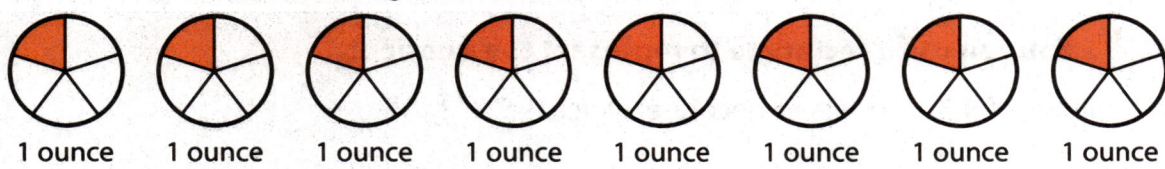

8 ounces ÷ 5 students = $\frac{8}{5}$ ounces per student

Another way: Think of 8 ounces as 5 ounces + 3 ounces. Each student will receive 1 full ounce of paint because there are 5 students. This leaves 3 ounces to divide.

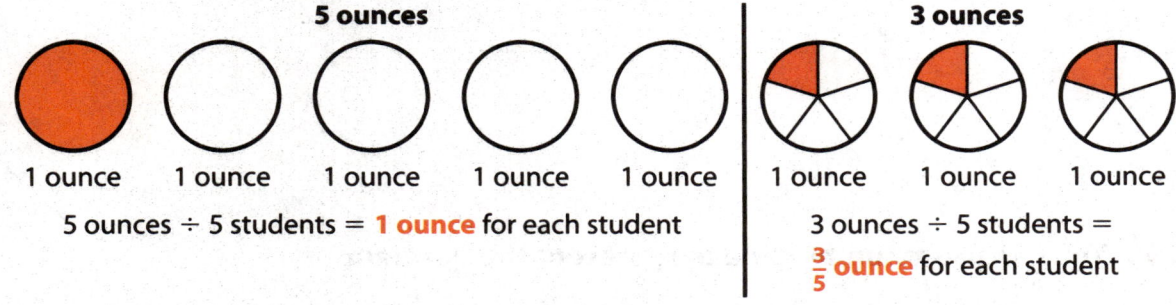

5 ounces | **3 ounces**

1 ounce 1 ounce 1 ounce 1 ounce 1 ounce | 1 ounce 1 ounce 1 ounce

5 ounces ÷ 5 students = **1 ounce** for each student

3 ounces ÷ 5 students = $\frac{3}{5}$ **ounce** for each student

Each student will receive 1 ounce plus $\frac{3}{5}$ ounce of paint.

8 ounces ÷ 5 students = **1$\frac{3}{5}$** ounces per student

▶ Reflect

1 Describe in words a problem that could be represented by the quotient $\frac{2}{5}$. Use the situation about sharing paint among art students.

_____

_____

_____

Lesson 12 Fractions as Division

Lesson 12 · Modeled and Guided Instruction

Learn About: Finding Fraction Quotients

Read the problem below. Then explore different ways to understand fractions as quotients.

> Jared, Monica, and Heather have 5 hallways to decorate for the student council. If they share the work equally, how much will each student decorate?

▶ Picture It You can use a rectangle to represent the quotient.

There are 5 hallways for 3 students to decorate, which is 5 ÷ 3.

If they share the work equally, each student can decorate $\frac{1}{3}$ of each hallway.

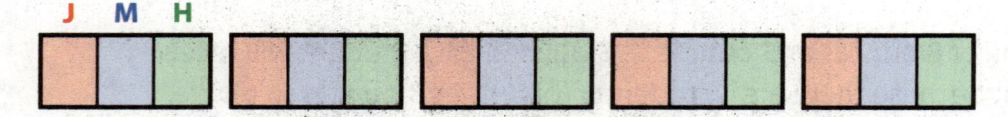

$\frac{1}{3} \times 5 = \frac{5}{3}$

$5 \div 3 = \frac{5}{3}$

▶ Model It You can use a number line to represent the quotient.

The number line is numbered from 0 to 5 because there are 5 hallways. It is divided into thirds because each student can decorate one third of each hallway.

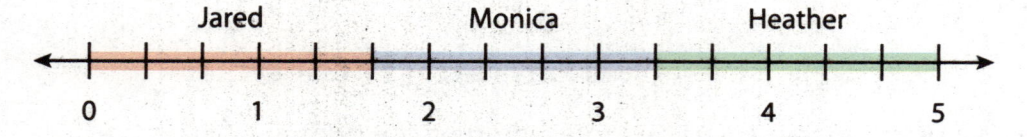

5 divided into 3 equal sections is five thirds, or $\frac{5}{3}$.

126 Lesson 12 Fractions as Division

Connect It Now you will solve the problem from the previous page by thinking about division as equal sharing.

2 How many thirds of a hallway are there to decorate in 5 hallways? _____ thirds

3 How many thirds of a hallway will each student decorate if the thirds are shared equally? _____ thirds

Write this as a fraction. _____ of a hallway

Here is another way to think of the problem.

4 How many whole hallways can each student decorate? _____
How many hallways remain after these are done? _____
If 5 hallways are shared equally by 3 students, there will be _____ whole hallway per student and _____ hallways remaining.

5 How much of 2 remaining hallways will each student decorate? _____

5 hallways ÷ 3 students = _____ whole and _____ hallway per student

Write this as a mixed number. _____ hallways

6 Write two division equations, one using the fraction and one using the mixed number, to represent the quotient.

_____

7 How does the bar in a fraction represent division? _____

_____

Try It Use what you just learned about fractions as quotients to solve these problems. Show your work on a separate sheet of paper.

8 Five friends are equally sharing 3 packs of football cards. How much of a pack will each friend get? _____

9 Mary made 18 ounces of apple chips. She evenly portions the apple chips between 4 containers. How many ounces of apple chips are in 1 container? Write a division expression to represent the problem and solve. _____

Lesson 12 Fractions as Division **127**

Lesson 12   Modeled and Guided Instruction

Learn About: Solving Word Problems with Fraction Quotients

Read the problem below. Then explore different ways to solve word problems with fractions as quotients.

> The 8 members of the math club are having lunch during a math competition. There are 12 sandwiches on the sandwich tray. If the club members share the sandwiches equally, how much will each member get?

▶ **Picture It**   You can draw a picture to represent the problem and quotient.

There are 12 sandwiches for the club members to share. The sandwiches make up a set.

12 sandwiches

Since there are 8 math club members sharing the sandwiches equally, draw lines to divide the set into 8 equal parts. Each club member gets $\frac{1}{8}$ of the set.

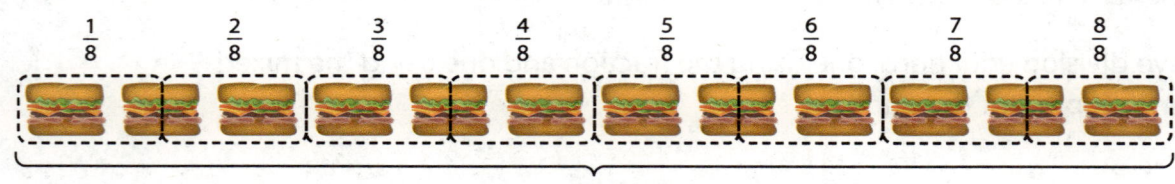

$\frac{1}{8}$ $\frac{2}{8}$ $\frac{3}{8}$ $\frac{4}{8}$ $\frac{5}{8}$ $\frac{6}{8}$ $\frac{7}{8}$ $\frac{8}{8}$

12 sandwiches

$\frac{1}{8} \times 12 = 12 \div 8$

▶ **Model It**   You can use an area model to show division.

There are 12 sandwiches for 8 club members to share, which is $12 \div 8$.

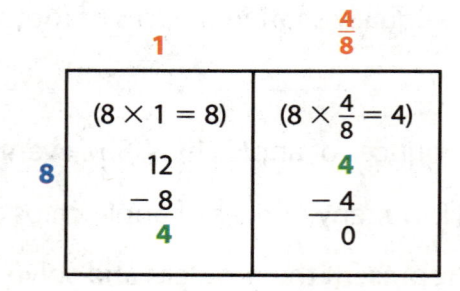

	?
8	12

	1	$\frac{4}{8}$
8	$(8 \times 1 = 8)$ 12 $-\ 8$ 4	$(8 \times \frac{4}{8} = 4)$ 4 $-\ 4$ 0

128   Lesson 12 Fractions as Division   ©Curriculum Associates, LLC   Copying is not permitted.

Connect It Now you will solve the problem from the previous page by thinking about division as equal parts of a set.

10 Look at *Picture It*. How many whole sandwiches can each of the 8 club members have? _____

How many sandwiches are divided? _____

How much of the remaining sandwiches does each member get? Express your answer as a fraction with an 8 in the denominator. _____

11 How many eighths of a sandwich will each club member get? _____ eighths

Write this as a fraction. _____ of the sandwiches

12 Look at *Model It*. 12 sandwiches ÷ 8 club members = _____ whole and _____ sandwich per club member.

Write this as a mixed number. _____ sandwiches

13 Write a division equation using a fraction to represent the quotient. Then write a division equation using a mixed number to represent the quotient.

_____

14 How does *Picture It* represent the division equation and quotient? Explain.

_____
_____

Try It Use what you just learned about solving problems with fractions as quotients to solve these problems.

15 Eight friends equally share a bag of 26 apples. How many apples will each friend get? Write a division expression to represent the problem and solve.

_____

16 Wyatt has 11 pounds of flour that he wants to split into 4 equally sized canisters. How many pounds of flour should he put in each canister? _____

Lesson 12 Fractions as Division   129

Lesson 12 Guided Practice

Practice: Fractions as Division

Study the example below. Then solve problems 17–19.

Example

Luke, Carter, and Ava have 2 quarts of juice. They want to share it evenly. How many quarts of juice will each of them get?

Look at how you could show your work using a model.

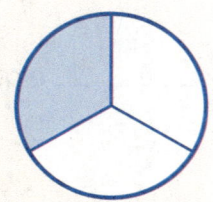

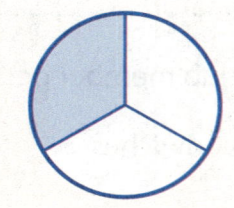

$$\frac{1}{3} \times 2 = \frac{2}{3}$$

Solution ___$\frac{2}{3}$ quart___

> 2 quarts are shared by 3 friends, so I know that each friend will have less than 1 quart of juice. That means the quotient is a fraction.

> **Pair/Share**
> Model the problem for 3 quarts of juice divided evenly among Luke, Carter, Ava, and Ava's little brother.

17 Erica has 7 square feet of garden space to plant carrots, beans, peppers, and lettuce. Suppose she gives each vegetable an equal amount of space. How much space will each vegetable get?

Show your work.

> Each vegetable will get at least 1 square foot of garden space. How will the rest of the space be divided up?

> **Pair/Share**
> What are some ways you can check your solution?

Solution _____

130 Lesson 12 Fractions as Division

©Curriculum Associates, LLC Copying is not permitted.

18 Daniel needs to make 6 pizzas. He has 123 pepperoni slices. He wants to use the same amount of pepperoni slices for each pizza. The number of pepperoni slices on each pizza should fall between what two whole numbers?

Show your work.

> How many whole pepperoni slices will each pizza get? What will happen with the remaining slices?

> **Pair/Share**
> Create a different division story to represent $\frac{123}{6}$.

Solution _____

19 Jonas is doing a science experiment with his class. The teacher has 21 fluid ounces of pond water to share equally with 10 pairs of students. How much pond water will Jonas and his science partner receive? Circle the letter of the correct answer.

A $\frac{10}{21}$ fluid ounce

B $1\frac{1}{10}$ fluid ounces

C 2 fluid ounces

D $\frac{21}{10}$ fluid ounces

Olivia chose **A** as the correct answer. How did she get that answer?

_____
_____
_____
_____

> About how much water will each pair of students receive? Will it be more or less than 2 fluid ounces?

> **Pair/Share**
> Does Olivia's answer make sense?

Lesson 12 Fractions as Division

Lesson 12 ▪ Independent Practice

Practice ▶ Fractions as Division

Solve the problems.

1. Teddy makes 50 fluid ounces of hot cocoa. He pours equal amounts of cocoa into 12 cups. The amount of hot cocoa in each cup will fall between which two amounts?

   A 2 and 3 fluid ounces

   B 3 and 4 fluid ounces

   C 4 and 5 fluid ounces

   D 5 and 6 fluid ounces

2. Pierce swims 10 laps in a pool in 8 minutes. He spent the same amount of time on each lap. How much time did each lap take him?

   A $\frac{2}{10}$ minute

   B $\frac{8}{10}$ minute

   C $\frac{10}{8}$ minutes

   D $1\frac{2}{8}$ minutes

3. Dani needs 8 equal sections from a board that is 13 meters long. Does the expression represent the largest possible length of 1 section of the board, in meters? Select *Yes* or *No* for each expression.

   a. $1\frac{5}{8}$ ☐ Yes ☐ No

   b. $\frac{8}{13}$ ☐ Yes ☐ No

   c. $\frac{13}{8}$ ☐ Yes ☐ No

   d. $8 \div 13$ ☐ Yes ☐ No

   e. $13 \times \frac{1}{8}$ ☐ Yes ☐ No

4 Which situation is represented by $\frac{25}{6}$? Circle the letter for all that apply.

   A Melanie equally shares 25 meters of paper to create 6 banners.

   B Quill gives away 6 baseball cards from a pack of 25 cards.

   C George invites 25 kids and 6 adults to his birthday party.

   D Becca creates 6 rows with 25 buttons each.

   E Joe makes 6 equal servings from a 25-ounce bag of peanuts.

5 Paco is trying to explain to his friend that $7 \div 2 = \frac{7}{2}$.

   Part A Draw a model or number line showing $7 \div 2 = \frac{7}{2}$.

   Part B Explain the equivalence of $7 \div 2$ and $\frac{7}{2}$ using words.

   _____

   _____

   _____

   _____

   Part C Describe in words a situation that can be represented by $7 \div 2$ or $\frac{7}{2}$.

   _____

   _____

   _____

✓ Self Check Go back and see what you can check off on the Self Check on page 103.

Lesson 13   Introduction
Understand Products of Fractions

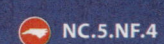

NC.5.NF.4

Think It Through

What does it mean to multiply a fraction by a whole number?

The ruler below shows three $\frac{1}{2}$-inch segments. You can represent this as $3 \times \frac{1}{2}$.

You have learned different ways to think about fractions and also about multiplication. Below are other ways to think about $3 \times \frac{1}{2}$ inch.

- Three $\frac{1}{2}$-inch sections is $\frac{3}{2}$ inches.
- Three $\frac{1}{2}$-inch sections is 1 whole inch plus $\frac{1}{2}$ inch, or $1\frac{1}{2}$ inches.
- $\frac{3}{2}$ inches is 3 times as long as $\frac{1}{2}$ inch.

Think   What does it mean to multiply a whole number by a fraction?

The ruler below shows a 3-inch segment divided into 2 equal parts. That's the same as multiplying 3 by $\frac{1}{2}$. You can represent this as $\frac{1}{2} \times 3$.

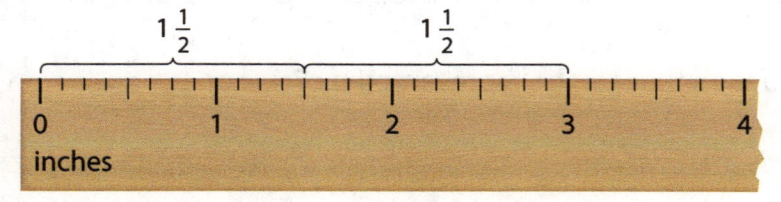

Below are ways to think about $\frac{1}{2} \times 3$.

- When 3 is divided into 2 equal parts, $\frac{1}{2} \times 3$ is one part.
- When 3 is divided into 2 equal parts, each part is $1\frac{1}{2}$ or $\frac{3}{2}$.
- $\frac{1}{2} \times 3 = 1\frac{1}{2}$

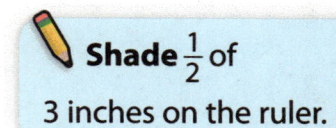

✏️ **Shade $\frac{1}{2}$ of 3 inches on the ruler.**

134   Lesson 13 Understand Products of Fractions

Think What happens when a fraction is multiplied by a fraction?

Find $\frac{1}{4}$ of a $\frac{1}{2}$-inch line segment, or $\frac{1}{4} \times \frac{1}{2}$.

The ruler below shows that $\frac{1}{2}$ inch is half of the whole 1 inch.

> If I say "find one fourth of," it means to divide by 4 or multiply by $\frac{1}{4}$.

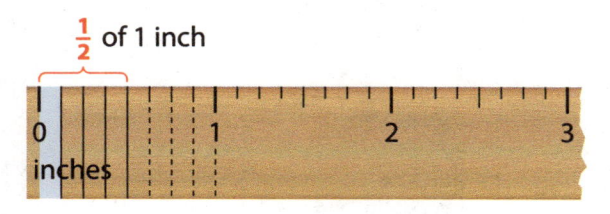

To find $\frac{1}{4}$ of $\frac{1}{2}$ inch, you divide the half inch into 4 equal parts. Remember that one-half inch is already part of the whole 1 inch. Divide the other half inch into 4 equal parts, and the whole will have 8 equal parts. So, $\frac{1}{4}$ of $\frac{1}{2}$ inch is 1 of 8 parts of the whole 1 inch: $\frac{1}{8}$ inch.

What if you need to find $\frac{3}{4}$ of a $\frac{1}{2}$-inch line segment? Above, you found that $\frac{1}{4}$ of $\frac{1}{2}$ inch is $\frac{1}{8}$ inch. Now you need to find three $\frac{1}{4}$s of $\frac{1}{2}$ inch.

$\frac{3}{4} \times \frac{1}{2}$

If $\frac{1}{4} \times \frac{1}{2}$ is $\frac{1}{8}$, then $\frac{3}{4}$ of $\frac{1}{2}$ is three $\frac{1}{8}$s, or $\frac{3}{8}$.

So, $\frac{3}{4} \times \frac{1}{2} = \frac{3}{8}$.

▶ Reflect

1. Based on the two problems above, what do you expect is the product of $\frac{2}{4} \times \frac{1}{2}$? Explain.

_____
_____
_____
_____

Lesson 13 Understand Products of Fractions **135**

Lesson 13 · Guided Instruction

Think About ▸ Multiplying a Fraction by a Fraction

🔍 **Let's Explore the Idea** Find the product of $\frac{3}{4} \times \frac{2}{3}$ with an area model.

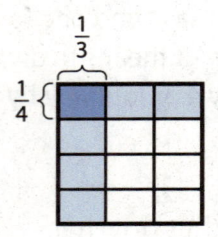

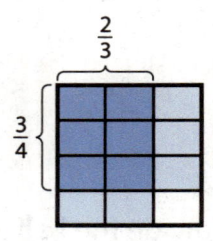

2 Look at the model on the left above.

Each column is what fraction of the whole? _____

Each row is what fraction of the whole? _____

3 The dark blue part shows $\frac{1}{4}$ of $\frac{1}{3}$.

How many parts are in the whole? _____

So, $\frac{1}{4} \times \frac{1}{3}$ is what fraction of the whole? _____

4 Look at the model on the right above.

Two columns are what fraction of the whole? _____

Three rows are what fraction of the whole? _____

5 The dark blue parts show $\frac{3}{4}$ of $\frac{2}{3}$.

What fraction of the whole is $\frac{3}{4} \times \frac{2}{3}$? _____

6 What is the product of $\frac{3}{4} \times \frac{2}{3}$? _____

136 Lesson 13 Understand Products of Fractions

Let's Talk About It
Complete these problems as a group.

7 Use the area model on the previous page to fill in the products in the table below.

$\frac{1}{4} \times \frac{1}{3} =$	$\frac{2}{4} \times \frac{1}{3} =$	$\frac{3}{4} \times \frac{1}{3} =$	$\frac{4}{4} \times \frac{1}{3} =$
$\frac{1}{4} \times \frac{2}{3} =$	$\frac{2}{4} \times \frac{2}{3} =$	$\frac{3}{4} \times \frac{2}{3} =$	$\frac{4}{4} \times \frac{2}{3} =$

8 Look at the numerators of each factor pair. How do these numbers relate to the numerator of the product? Now look at the denominators of each factor pair. How do these denominators relate to the denominator of the product?

_____

_____

_____

▶ **Try It Another Way** Work with your group to connect the model to the multiplication expression.

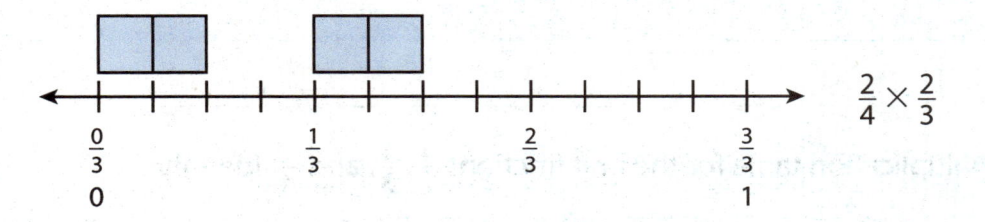

$\frac{2}{4} \times \frac{2}{3}$

Number line labeled: $\frac{0}{3}$ (0), $\frac{1}{3}$, $\frac{2}{3}$, $\frac{3}{3}$ (1)

9 Each third on the number line is divided into how many equal parts? _____

Each of these parts is what fraction of the whole? _____

10 How many fourths of the first $\frac{1}{3}$ are shaded? _____

How many fourths of the second $\frac{1}{3}$ are shaded? _____

11 How many twelfths of the whole are shaded? _____

12 Explain how the model shows $\frac{2}{4} \times \frac{2}{3}$. What is the product? _____

_____

_____

©Curriculum Associates, LLC   Copying is not permitted.   Lesson 13 *Understand* Products of Fractions   **137**

Lesson 13 Guided Practice
Connect: Ideas About Multiplying Fractions

Talk through these problems as a class. Then write your answers below.

13 Describe Tell what multiplication problem the model shows. Explain why. _____

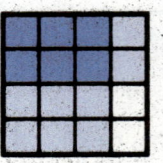

_____

_____

_____

_____

14 Explain Landon said that $\frac{2}{3} \times \frac{1}{3} = 1$. Tell how Landon found his answer, and then explain how to find the correct product. _____

_____

_____

_____

_____

15 Create Complete the multiplication table for the unit fractions $\frac{1}{2}$, $\frac{1}{3}$, and $\frac{1}{4}$. Identify two patterns in the table.

×	$\frac{1}{2}$	$\frac{1}{3}$	$\frac{1}{4}$
$\frac{1}{2}$			
$\frac{1}{3}$			
$\frac{1}{4}$			

_____

_____

138 Lesson 13 Understand Products of Fractions

Lesson 13 · Independent Practice

Apply ▶ Ideas About Multiplying Fractions

16 Put It Together Use what you have learned to complete this task.

Part A Start with a sheet of paper. Fold it in half, then fold it in half again. Continue folding the paper in half. As you do, fill in the blanks to tell what fraction of the whole each equal part is.

$\frac{1}{2} \times 1 =$ _____

$\frac{1}{2} \times \frac{1}{2} =$ _____

$\frac{1}{2} \times$ _____ $=$ _____

$\frac{1}{2} \times$ _____ $=$ _____

Part B Write the products as the terms of a pattern. Then look at the denominators of the fractions. Each time you multiply by $\frac{1}{2}$, what happens to the denominator?

_____
_____
_____

Without folding the paper, what do you think the next fraction in the pattern will be? Why?

_____
_____
_____

Part C Now look at the sizes of the equal parts of the folded paper. Each time you fold the paper in half, what happens to the size of the parts?

_____
_____

Lesson 14   Introduction
Multiply Fractions Using an Area Model

Use What You Know

You have learned about multiplying fractions. Now you will use area models to multiply fractions. Take a look at this problem.

> Mr. Thompson has a 1-meter-square whiteboard. He creates a square with $\frac{2}{4}$-meter sides on his whiteboard to post a weekly puzzle. How many square meters of whiteboard space does he use for the weekly puzzle?

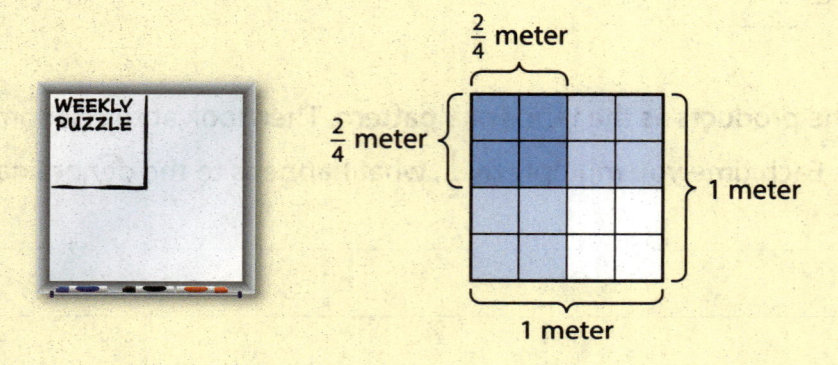

a. Write a multiplication equation for the area of the square with side lengths of 1 meter.

_____

b. What expression can you write for the area of the smaller square with side lengths of $\frac{2}{4}$ meter? _____

c. The square with the darker shading is 2 units wide and 2 units long. How many square units is this? _____

d. What is the total number of small squares in the large square? _____

e. What fraction of the whole square is the part with the darker shading? _____

f. The area of the square with the darker shading is _____ square meter.

140   Lesson 14 Multiply Fractions Using an Area Model

▶▶ Find Out More

The area of a square with sides $\frac{2}{4}$ meter represents the product $\frac{2}{4} \times \frac{2}{4}$.

- The length of each unit is $\frac{1}{4}$ meter. The length and the width of the blue square are both 2 units, or $\frac{2}{4}$ meter.
- The area of one square unit is $\frac{1}{16}$ square meter. The area of the blue square is 4 square units, or $\frac{4}{16}$ square meter.

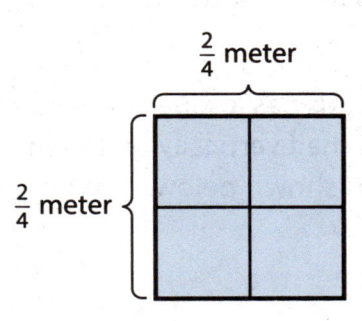

$\frac{2}{4}$ meter $\times \frac{2}{4}$ meter $= \frac{4}{16}$ square meter

You can also think about multiplying $\frac{2}{4} \times \frac{2}{4}$ another way.

$\frac{2}{4}$ is $2 \times \frac{1}{4}$.

Rewrite. $\quad\quad\quad\quad\quad\quad\quad\quad \frac{2}{4} \times \frac{2}{4} = \left(2 \times \frac{1}{4}\right) \times \left(2 \times \frac{1}{4}\right)$

Change the order of the factors. $\quad\quad = 2 \times 2 \times \frac{1}{4} \times \frac{1}{4}$

Then multiply. $\quad\quad\quad\quad\quad\quad = 4 \times \frac{1}{16}$, or $\frac{4}{16}$

▶ Reflect

1 Look at the models on these pages. Explain why multiplying the numerators tells you the number of parts in the product.

_____

_____

_____

_____

Lesson 14 Multiply Fractions Using an Area Model

Lesson 14 Modeled and Guided Instruction

Learn About Multiplying Unit Fractions to Find Areas

Read the problem below. Then explore different ways to understand multiplying two unit fractions.

> Titus has a square sheet of paper measuring 1 foot on each side. He folds the paper in half vertically and then folds it into fourths horizontally. Titus unfolds the paper as shown below. What is the area of each part?

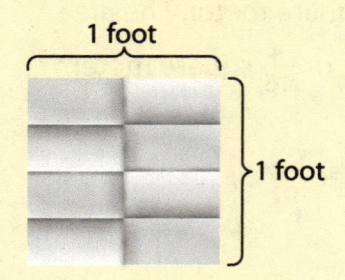

▶ Picture It You can understand the problem by picturing parts of the unfolded paper.

The folding created 8 equal sections.

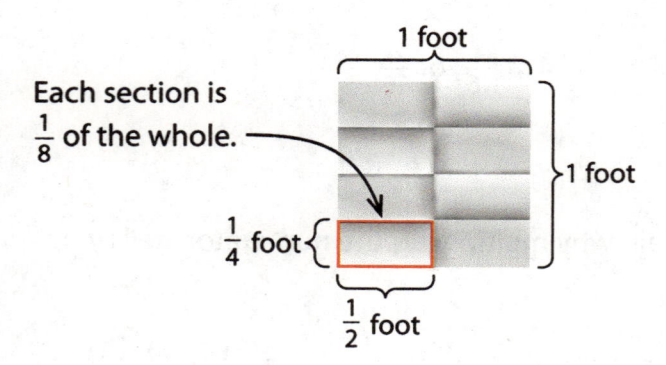

Each section is $\frac{1}{8}$ of the whole.

Each section is $\frac{1}{2}$ foot long and $\frac{1}{4}$ foot wide.

▶ Model It You can model the problem with an equation.

Each equal part of the paper has a length of $\frac{1}{2}$ foot and width of $\frac{1}{4}$ foot.

area = $\frac{1}{2}$ foot × $\frac{1}{4}$ foot

$\frac{1}{2}$ foot × $\frac{1}{4}$ foot = $\frac{1 \times 1}{2 \times 4}$ square foot

Lesson 14 Multiply Fractions Using an Area Model   ©Curriculum Associates, LLC   Copying is not permitted.

Connect It Now you will solve the problem from the previous page by connecting the model to the equation.

2 Explain why each part of the paper is $\frac{1}{2}$ foot long. _____

3 Explain why each part of the paper is $\frac{1}{4}$ foot wide. _____

4 What expression do you use to find the area of 1 part of the paper? _____

5 Multiply the denominators of these fractions. How does the product relate to the size of the units compared to the size of the whole model? _____

6 Multiply the numerators of these fractions. How does the product relate to the number of outlined parts of the model? _____

7 Explain how to find the area of one part of the sheet of paper.

Try It Use what you just learned about multiplying unit fractions to solve this problem. Show your work on a separate sheet of paper.

8 What is the area of a paper strip with a width of $\frac{1}{3}$ yard and length of $\frac{1}{6}$ yard?

Lesson 14 Multiply Fractions Using an Area Model

Lesson 14   Modeled and Guided Instruction

Learn About Multiplying Fractions Greater than One

Read the problem below. Then explore different ways to understand multiplying mixed numbers and fractions greater than 1.

> A postage stamp has a width of $\frac{3}{4}$ inch and length of $\frac{3}{2}$ inches. What is the area of the stamp in square inches?

▶ **Picture It** You can picture the problem using the area of a rectangle.

The first area model shows $\frac{1}{4}$ inch $\times \frac{1}{2}$ inch $= \frac{1}{8}$ square inch. The second model uses the same $\frac{1}{8}$-square-inch parts to show an area that is $\frac{3}{4}$ inch $\times \frac{3}{2}$ inches.

$\frac{1}{4} \times \frac{1}{2}$ $\frac{3}{4} \times \frac{3}{2}$

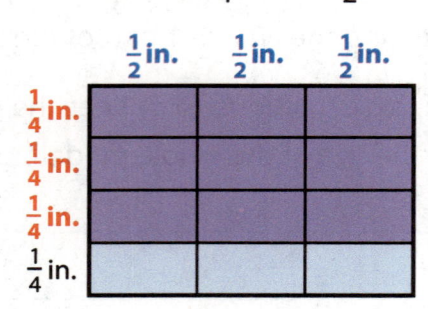

▶ **Model It** You can model the problem with an equation.

The dimensions of the stamp are $\frac{3}{4}$ inch and $\frac{3}{2}$ inches, so multiply the fractions to find the area.

$$\text{area} = \frac{3}{4} \times \frac{3}{2} = \frac{3 \times 3}{4 \times 2}$$

144   Lesson 14 Multiply Fractions Using an Area Model

Connect It Now you will solve the problem from the previous page by connecting the model to the equation.

9 Look at the area model for $\frac{3}{4} \times \frac{3}{2}$. Explain why each part shows $\frac{1}{8}$ square inch.

_____

_____

10 What is the total width of the three $\frac{1}{2}$-inch columns? _____

What is the total height of the three $\frac{1}{4}$-inch rows? _____

11 How many $\frac{1}{8}$-square-inch parts are shaded purple? _____ parts

Write the area of the purple section as a fraction greater than 1. _____ square inches

12 Now look at the equation in *Model It*. Multiply the numerators, multiply the denominators, and write the fraction. How does this product compare with the one shown by the area model?

_____

_____

_____

13 How do you multiply fractions greater than 1?

_____

_____

_____

_____

Try It Use what you just learned about multiplying fractions to solve this problem. Show your work on a separate sheet of paper.

14 Bernice's math workbook is $\frac{2}{3}$ foot wide and $\frac{5}{6}$ foot long. What is the area of a page in the workbook? _____

Lesson 14 Multiply Fractions Using an Area Model

Lesson 14 — Guided Practice

Practice: Multiplying Fractions to Find Area

Study the example below. Then solve problems 15–17.

Example

Rachel is designing a newspaper ad. The ad will include a piece of art whose dimensions are $\frac{3}{4}$ inch long and $\frac{1}{2}$ inch wide. How many square inches of space will the art cover?

Look at how you could show your work using an area model.

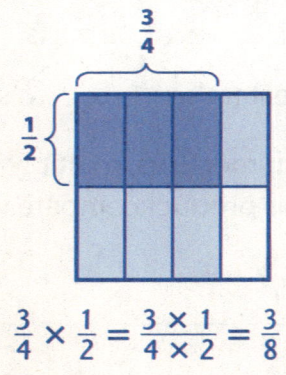

$$\frac{3}{4} \times \frac{1}{2} = \frac{3 \times 1}{4 \times 2} = \frac{3}{8}$$

Solution $\frac{3}{8}$ square inch

> $\frac{1}{4}$ inch × $\frac{1}{2}$ inch is $\frac{1}{8}$ square inch. How many eighth square inches are shown in the model?

> **Pair/Share**
> How can you write $\frac{3}{4} \times \frac{1}{2}$ as a product of unit fractions and whole numbers?

---

15 What is the area of a rectangle with a length of $\frac{1}{2}$ yard and a width of $\frac{11}{4}$ yards? Write an equation to represent your solution.
Show your work.

> How can I represent a fractional side length with an area model?

> **Pair/Share**
> Find the area of a rectangle with side lengths of $\frac{3}{4}$ yard and $\frac{6}{3}$ yards. How is the model different?

Solution _____

146 Lesson 14 Multiply Fractions Using an Area Model

16 Brent is designing a poster that has an area of 1 square foot. He is going to paste a photo collage on a section of the poster that is $\frac{1}{3}$ foot wide and $\frac{3}{4}$ foot long. What part of a square foot will the photo collage cover?

Show your work.

If I draw a square to represent a square foot, how can I represent thirds and fourths on the square?

Solution _____

Pair/Share
Write an equation to represent your model. Explain the meaning of the numerators.

---

17 What is the area of the square?
Circle the letter of the correct answer.

$\frac{6}{7}$ yd

$\frac{6}{7}$ yd

A $\frac{36}{49}$ square yard

B $\frac{12}{14}$ square yard

C $\frac{49}{36}$ square yards

D $\frac{12}{7}$ square yards

Ollie chose **D** as the correct answer. How did he get that answer?

Think about the size of the two fractions. Will the product of the fractions be greater than 1 or less than 1?

Pair/Share
Does Ollie's answer make sense?

Lesson 14 Multiply Fractions Using an Area Model   147

Lesson 14 👤 **Independent Practice**

Practice ▶ Multiplying Fractions to Find Area

Solve the problems.

1 The square below represents 1 square unit.

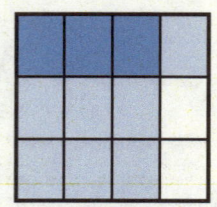

Which expression represents the area of the dark blue section?

A $\frac{4}{3} \times \frac{3}{1}$ square units

B $\frac{3}{4} \times \frac{1}{3}$ square units

C $\frac{1}{4} \times \frac{1}{3}$ square units

D $\frac{4}{3} \times \frac{1}{3}$ square units

2 Fill in the missing numbers to make the equation true. Then complete the area model to check your answer.

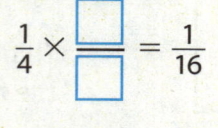

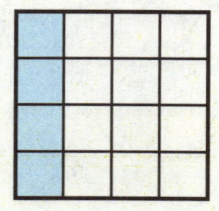

$$\frac{1}{4} \times \frac{\square}{\square} = \frac{1}{16}$$

3 Which products could you find by shading the model below? Circle the letter for all that apply.

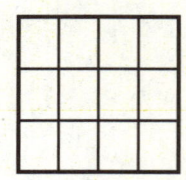

A $\frac{3}{4} \times \frac{1}{3}$

B $\frac{1}{3} \times \frac{1}{6}$

C $\frac{2}{3} \times \frac{1}{4}$

D $\frac{5}{3} \times \frac{1}{4}$

E $\frac{3}{4} \times \frac{3}{4}$

148 Lesson 14 Multiply Fractions Using an Area Model

4 Draw an area model to represent the expression $\frac{8}{4}$ inches × $\frac{2}{4}$ inch.

5 Explain how to find the area of the model you drew in problem 4. Then find the area.

6 Write the dimensions of a different rectangle that has the same area as the rectangle you drew in problem 4. Show how you know the area is the same.

Self Check Go back and see what you can check off on the Self Check on page 103.

Lesson 15   Introduction
Understand Multiplication as Scaling

NC.5.NF.4

Think It Through

What does scaling mean?

Think of how you use words and phrases such as "double," "triple," "half of," or "take one tenth." These words and phrases describe **scaling,** or changing the size of a quantity. *Stretching* and *shrinking* are two different ways to scale a quantity.

The table below shows some ways that a quantity of 6 can be scaled.

	Words	Symbols
stretching	6 **doubled** is 12.	**2** × 6 = 12
	6 **tripled** is 18.	**3** × 6 = 18
shrinking	**Half of** 6 is 3.	$\frac{1}{2}$ × 6 = 3
	A tenth of 6 is $\frac{6}{10}$.	$\frac{1}{10}$ × 6 = $\frac{6}{10}$

Think How can you use models to show what scaling means?

Below is a rectangle with an area of 6 square units.

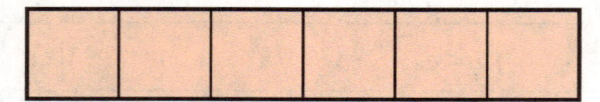

The model for 2 × 6 has an area that is double the size of the original rectangle.

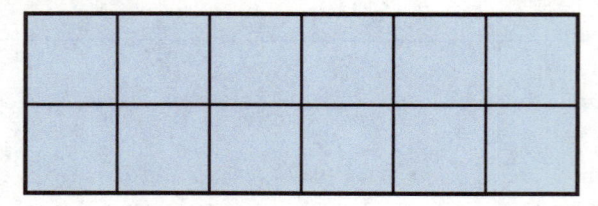

The model for $\frac{1}{2}$ × 6 has an area that is half the size of the original rectangle.

Circle the factor that describes how the rectangle is being stretched or shrunk.

> **Think** How does the size of the factors affect the product?

Products aren't always greater than their factors. The table below shows products of different factors multiplied by 6.

Factor	$\frac{1}{10}$	$\frac{1}{3}$	$\frac{1}{2}$	$\frac{5}{6}$	1	$\frac{4}{3}$	2	$2\frac{1}{2}$	3
Product of factor and 6	$\frac{6}{10}$	2	3	5	6	8	12	15	18

Look at the products that are less than 6, then look at those that are greater than 6. What do you notice about the factors?

Notice that the products are sometimes less than 6, sometimes greater than 6, and sometimes equal to 6.

What do the **products less than 6** have in common? The other factor is less than 1.

What do the **products greater than 6** have in common? The other factor is greater than 1.

- If you multiply 6 by a factor less than 1, the product will be less than 6.
- If you multiply 6 by a factor greater than 1, the product will be greater than 6.
- If you multiply 6 by 1, or a factor equivalent to 1, the product will be 6.

▶ **Reflect**

1 Describe the products you get if you multiply 8 by factors less than 1. Describe the products you get if you multiply 8 by factors greater than 1. Give some examples that justify your answers.

_____
_____
_____
_____
_____

Lesson 15 Understand Multiplication as Scaling 151

Lesson 15　Guided Instruction

Think About ▸ Comparing Factors and Products

🔍 **Let's Explore the Idea** A number line can help you see what happens when a fraction is multiplied by a factor less than 1.

2 You can show $\frac{1}{3} \times \frac{3}{4}$ on a number line. If you break up $\frac{3}{4}$ into 3 equal parts, each part is $\frac{1}{4}$.

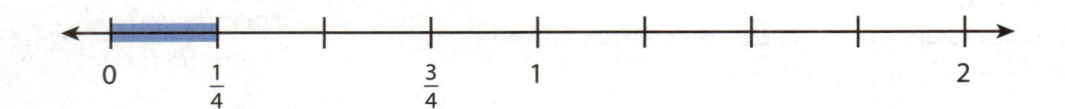

$$\frac{1}{3} \times \frac{3}{4} = \underline{}$$

Is the product *less than, greater than,* or *equal to* $\frac{3}{4}$? _____

3 Show $\frac{2}{3} \times \frac{3}{4}$ on a number line. If you break up $\frac{3}{4}$ into 3 equal parts, each part is $\frac{1}{4}$. Since you multiply by $\frac{2}{3}$, you need 2 of those parts. Shade and label $\frac{2}{3}$ of $\frac{3}{4}$.

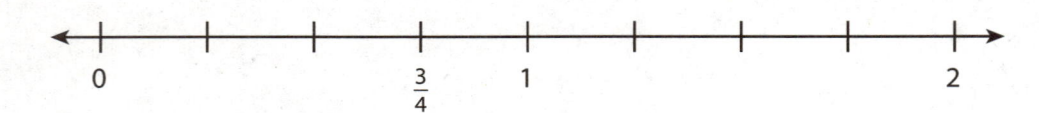

$$\frac{2}{3} \times \frac{3}{4} = \underline{}$$

Is the product *less than, greater than,* or *equal to* $\frac{3}{4}$? _____

4 You multiplied $\frac{3}{4}$ by two different factors. How do both factors relate to the number 1? What happens when you multiply a given fraction by a factor less than 1?

_____

_____

_____

Let's Talk About It A number line can also help you see what happens when a fraction is multiplied by a factor greater than 1.

5 Shade the number line to show $\frac{4}{3} \times \frac{3}{4}$.

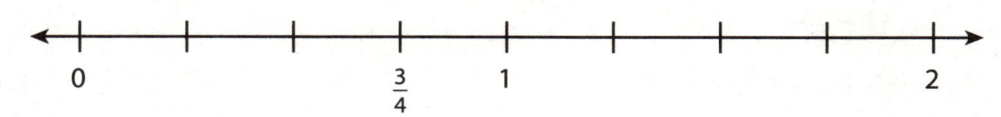

0 — $\frac{3}{4}$ — 1 — 2

$\frac{4}{3} \times \frac{3}{4} = $ _____

Is the product *less than, greater than,* or *equal to* $\frac{3}{4}$? _____

6 Shade and label the number line to show $\frac{7}{3} \times \frac{3}{4}$.

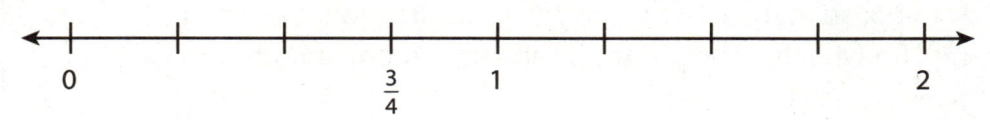

0 — $\frac{3}{4}$ — 1 — 2

$\frac{7}{3} \times \frac{3}{4} = $ _____

Is the product *less than, greater than,* or *equal to* $\frac{3}{4}$? _____

7 Think about how each of your answers compared to $\frac{3}{4}$. What can you say about the product of a given fraction and a factor greater than 1?

_____

Try It Another Way Explore multiplying $\frac{3}{4}$ by a fraction using an area model.

The model to the right represents $\frac{3}{4}$.

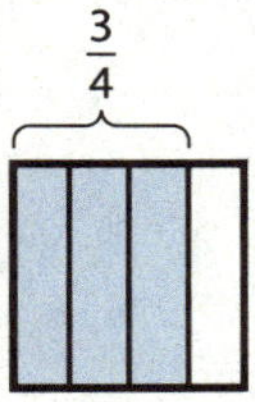

8 Show $\frac{1}{2} \times \frac{3}{4}$ using the area model.

9 $\frac{1}{2} \times \frac{3}{4} = $ _____

10 Is the product *less than, greater than,* or *equal to* $\frac{3}{4}$? _____

11 Could you have answered problem 10 without drawing a model? Explain.

_____

Lesson 15 Understand Multiplication as Scaling

Lesson 15 Guided Practice

Connect: Ideas About Factors and Products

Talk through these problems as a class, then write your answers below.

12 Analyze Use reasoning to order the following expressions from least to greatest. Don't calculate any of the products. Explain your reasoning.

$\frac{7}{9} \times 348{,}980$ $\qquad$ $\frac{12}{11} \times 348{,}980$ $\qquad$ $\frac{50}{50} \times 348{,}980$

_____
_____
_____
_____

13 Explain Gillian said that the product of a given number and a fraction is always less than the given number. Explain what is wrong with Gillian's statement and give an example that does not follow her rule.

_____
_____
_____
_____

14 Compare Represent the expression $\frac{4}{4} \times \frac{8}{3}$ with a model. Write a sentence comparing the product with $\frac{8}{3}$. Explain your reasoning.

_____
_____
_____

Lesson 15 Independent Practice

Apply ▶ Ideas About Factors and Products

15 Put It Together You can compare the size of a product to the size of the factors in a multiplication equation if you know whether the factors are greater than, less than, or equal to 1.

Part A Write a multiplication equation (different from any in this lesson) in which the product is greater than both of the factors. Draw a model to support your answer.

_____

_____

Part B Write a multiplication equation (different from any in this lesson) in which both the factors are fractions and the product is less than both of the factors. Draw a model to support your answer.

_____

_____

Part C Write a multiplication equation (different from any in this lesson) in which the product is equal to one of the factors.

_____

_____

Lesson 16 Introduction
Multiply Fractions in Word Problems

NC.5.NF.4

Use What You Know

Now that you have learned how to multiply fractions, take a look at this problem.

Grayson lives $\frac{2}{3}$ mile from the park. He has already walked $\frac{1}{2}$ of the way to the park. How far has Grayson walked?

a. You can draw a model to help you solve the problem. Locate a point on the number line below to show what fraction of a mile Grayson lives from the park.

b. Label the point to show the distance to the park.

c. Shade the segment that shows one half of the distance to the park.

d. One half of the distance to the park is _____ mile. Label this distance.

e. Explain how you can use the model to show how far Grayson has already walked.

156 Lesson 16 Multiply Fractions in Word Problems

▶▶ Find Out More

Consider the problem from the previous page. The distance you need to find is a fraction of a fraction: $\frac{1}{2}$ of $\frac{2}{3}$ mile. On the previous page you used a model to find this distance. You can also use an equation.

Finding $\frac{1}{2}$ of a number is the same as multiplying the number by $\frac{1}{2}$.

$\frac{1}{2}$ of $\frac{2}{3}$ means $\frac{1}{2} \times \frac{2}{3}$.

To multiply two fractions, multiply the numerators to get the numerator of the product, and multiply the denominators to get the denominator of the product.

$$\frac{1}{2} \times \frac{2}{3} = \frac{1 \times 2}{2 \times 3} = \frac{2}{6}$$

The fraction $\frac{2}{6}$ is equivalent to $\frac{1}{3}$. To find equivalent fractions, multiply or divide the numerator and the denominator of the fraction by the same number.

$$\frac{2}{6} = \frac{2 \div 2}{6 \div 2} = \frac{1}{3}$$

Grayson has walked $\frac{1}{3}$ mile.

The product, $\frac{2}{6}$, is less than $\frac{2}{3}$, so it is reasonable. When you multiply $\frac{2}{3}$ by a factor less than 1, the product should be less than $\frac{2}{3}$.

▶ Reflect

1 Which method, *drawing a model* or *writing an equation*, made more sense to you for solving this problem? Why?

_____

_____

_____

_____

Lesson 16 Multiply Fractions in Word Problems

Lesson 16   Modeled and Guided Instruction

Learn About ▶ Multiplying Fractions in Word Problems

Read the problem below. Then explore different ways to understand how to find a fraction of a fraction.

> Brandon's mother left $\frac{3}{4}$ of a pizza on the counter. If Brandon eats $\frac{2}{3}$ of it, how much of the whole pizza did Brandon eat?

▶ **Picture It** You can draw a picture to help you understand the problem.

Show $\frac{3}{4}$ of a pizza.

Since Brandon eats $\frac{2}{3}$ of what is left, outline 2 of the 3 pieces that are left. You can see from the outlined parts how much of the whole pizza Brandon ate.

▶ **Model It** You can write an equation to help you understand the problem.

You need to find a fraction of a fraction: $\frac{2}{3}$ of $\frac{3}{4}$ of a pizza.

$\frac{2}{3}$ of $\frac{3}{4}$ means $\frac{2}{3} \times \frac{3}{4}$.

$$\frac{2}{3} \times \frac{3}{4} = \frac{2 \times 3}{3 \times 4}$$

Lesson 16 Multiply Fractions in Word Problems

Connect It Now you will solve the problem from the previous page using both strategies.

2 Look at *Picture It*. Why did you outline 2 of the 3 parts of the pizza?

_____

3 How much of the whole pizza did Brandon eat? Explain your reasoning.

_____

_____

4 Look at *Model It*. How do you know that you should multiply $\frac{2}{3} \times \frac{3}{4}$?

_____

_____

5 What is $\frac{2 \times 3}{3 \times 4}$? _____

Is this product the same as your answer to problem 3? Explain.

_____

6 What strategies can you use to solve a word problem that involves multiplying fractions? _____

_____

_____

Try It Use what you just learned about finding products of fractions to solve these problems. Show your work on a separate sheet of paper.

7 Lewis walked $\frac{9}{10}$ of a mile. His friend walked $\frac{2}{5}$ of the way with him. How many miles did Lewis's friend walk with him? _____

8 Jamie worked $\frac{5}{6}$ hour filing papers for her mother. She listened to music for $\frac{2}{5}$ of the time she spent filing. How much time did Jamie spend listening to music?

_____

Lesson 16 Multiply Fractions in Word Problems

Lesson 16 — Modeled and Guided Instruction

Learn About: Multiplying Mixed Numbers in Word Problems

Read the problem below. Then explore different ways to understand multiplying fractions and mixed numbers.

> Janie has $2\frac{3}{4}$ yards of yellow ribbon. She uses $\frac{1}{2}$ of the ribbon to decorate bookmarks for her cousins. How many yards of ribbon did Janie use for the bookmarks?

▶ **Picture It** You can use an area model to help you understand the problem.

The purple shaded region of the area model shows **half** of $2\frac{3}{4}$.

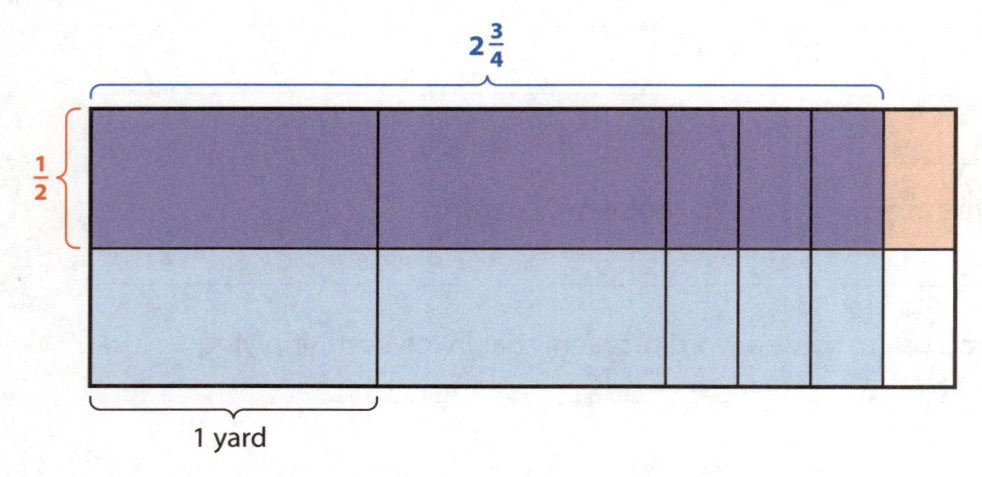

▶ **Model It** You can write an equation to help you understand the problem.

You can write $2\frac{3}{4}$ as a fraction.

$$2\frac{3}{4} = 2 + \frac{3}{4}$$
$$= \frac{8}{4} + \frac{3}{4}$$
$$= \frac{11}{4}$$

You need to find a fraction of a fraction: $\frac{1}{2}$ of $\frac{11}{4}$ yards of ribbon.

$\frac{1}{2}$ of $\frac{11}{4}$ means $\frac{1}{2} \times \frac{11}{4}$.

$$\frac{1}{2} \times \frac{11}{4} = \frac{1 \times 11}{2 \times 4}$$

160 Lesson 16 Multiply Fractions in Word Problems

Connect It Now you will solve the problem from the previous page comparing the two strategies.

9. Does Janie use more or less than $2\frac{3}{4}$ yards of ribbon for the bookmarks? Explain.

_____

_____

10. How many yards of ribbon did Janie use? _____ Explain how you can use the picture to answer the question. _____

_____

11. How do you know that you should multiply to solve the problem? _____

_____

_____

12. How can you multiply $2\frac{3}{4}$ by $\frac{1}{2}$? _____

_____

13. What is $\frac{1}{2} \times 2$? _____ What is $\frac{1}{2} \times \frac{3}{4}$? _____

    Add the two products. _____ + _____ = _____

    Is this result the same as your answer to problem 10? _____

14. Suppose Janie had $2\frac{1}{4}$ yards of ribbon and used $\frac{1}{2}$ of the ribbon for bookmarks. Explain how you could find how many yards of ribbon she used. _____

_____

_____

Try It Use what you just learned about multiplying fractions and mixed numbers to solve this problem. Show your work on a separate sheet of paper.

15. Izzy has $3\frac{1}{2}$ yards of rope. She uses $\frac{3}{5}$ of the rope to attach a tire swing to a tree in the yard. How many yards of rope does Izzy use for the tire swing? _____

Lesson 16 Guided Practice

Practice: Multiplying Fractions in Word Problems

Study the example below. Then solve problems 16–18.

Example

Chris is $4\frac{1}{4}$ feet tall. His mom is $1\frac{1}{2}$ times as tall as Chris. How tall is Chris's mom?

Look at how you can solve this problem using an equation.

$$4\frac{1}{4} \times 1 = 4\frac{1}{4}$$

$$4\frac{1}{4} \times \frac{1}{2} = \left(4 \times \frac{1}{2}\right) + \left(\frac{1}{4} \times \frac{1}{2}\right) = 2 + \frac{1}{8}$$

$$4\frac{1}{4} + 2 + \frac{1}{8} = 6\frac{1}{4} + \frac{1}{8} = 6\frac{2}{8} + \frac{1}{8} = 6\frac{3}{8}$$

Solution: $6\frac{3}{8}$ feet

The student wrote and solved an equation to solve the problem.

Pair/Share
How does the product compare to $4\frac{1}{4}$ feet?

16 Josh exercises at the gym $3\frac{3}{4}$ hours a week. He spends $\frac{2}{5}$ of his time at the gym lifting weights. How many hours a week does Josh spend lifting weights at the gym?

Show your work.

How do I know what operation to use to solve this problem?

Pair/Share
What is a reasonable estimate for the number of hours Josh lifts weights each week?

Solution _____

162 Lesson 16 Multiply Fractions in Word Problems

17 A field is in the shape of a rectangle $\frac{1}{4}$ mile long and $\frac{3}{4}$ mile wide. What is the area of the field?

Show your work.

What model can I use to help understand this problem?

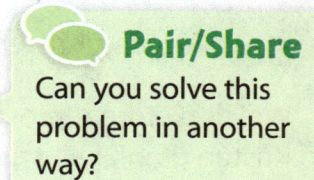

Pair/Share
Can you solve this problem in another way?

Solution _____

---

18 Ari had $\frac{3}{4}$ of a bag of popcorn. His friends ate $\frac{1}{2}$ of his popcorn. What fraction of the whole bag of popcorn did Ari's friends eat?

Circle the letter of the correct answer.

A $\frac{1}{4}$

B $\frac{3}{8}$

C $\frac{5}{4}$

D $\frac{3}{2}$

Kayla chose **A** as the correct answer. How did she get that answer?

_____

_____

_____

What equation can I write to solve this problem?

Pair/Share
Does Kayla's answer make sense?

Lesson 16 Multiply Fractions in Word Problems

Lesson 16 Independent Practice

Practice: Multiplying Fractions in Word Problems

Solve the problems.

1. On Sunday, Kristen bought a carton of 24 bottles of water.

   - On Monday, Kristen drank $\frac{1}{6}$ of the bottles in the carton.
   - On Tuesday, Kristen drank $\frac{1}{4}$ of the bottles that remained in the carton after Monday.

   Which picture represents the number of bottles of water remaining in the carton after Kristen drank her water on Tuesday?

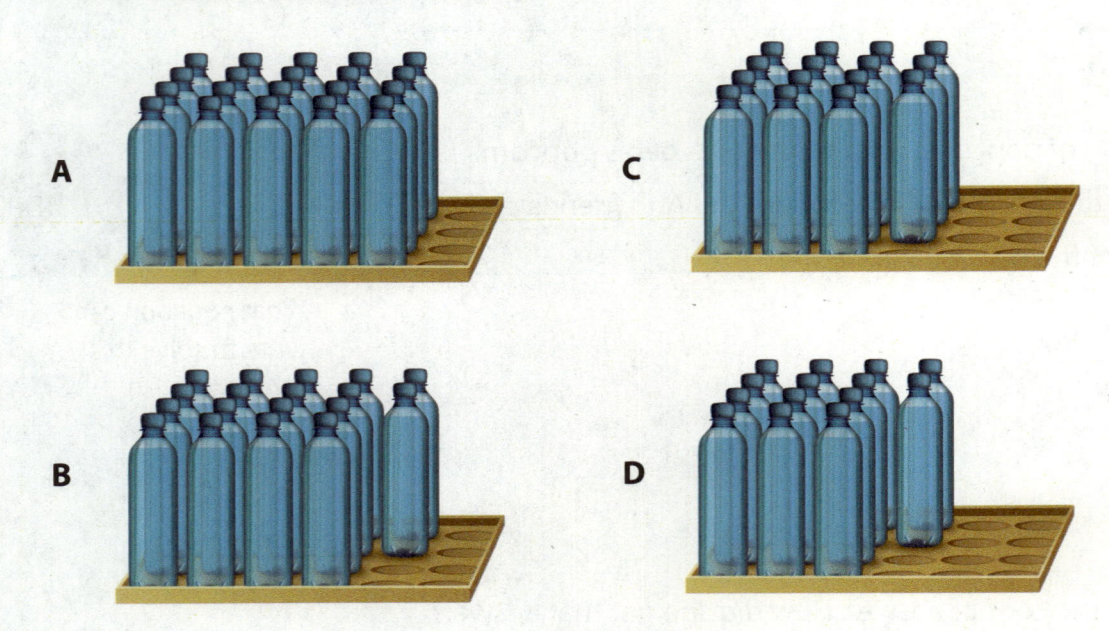

   A

   B

   C

   D

2. Milo's pancake recipe makes 9 servings. It calls for $\frac{3}{4}$ cup milk. Milo wants to make 6 servings. How much milk will he need?

   Answer _____ cup

3. Look at the rectangle below.

   $2\frac{2}{5}$ in.

   $4\frac{1}{8}$ in.

   What is the area of the rectangle? _____ square inches

4. Lily designed the letters of her name on the computer and printed them on paper. The table below shows the width and height of the printed letters.

Letter	Width	Height
L	$2\frac{1}{2}$ in.	4 in.
I	$1\frac{3}{4}$ in.	4 in.
Y	$1\frac{2}{3}$ in.	4 in.

   She used a copier to change the size of the letters to $\frac{3}{4}$ of the original size. Make a table to show the new dimensions of each letter.

   Show your work.

✓ Self Check Go back and see what you can check off on the Self Check on page 103.

Lesson 16 Multiply Fractions in Word Problems

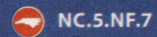

Lesson 17   Introduction
Understand Division with Unit Fractions

Think It Through

How is dividing with fractions related to multiplying with fractions?

You know that multiplication and division are related. Dividing 8 by 4, for example, gives the same result as multiplying 8 by $\frac{1}{4}$.

$8 \div 4 = 2$

$8 \times \frac{1}{4} = 2$

Dividing with unit fractions works the same way. You can solve a division problem involving fractions by multiplication.

Think What does dividing a unit fraction by a whole number mean?

Mrs. Cook wants to share $\frac{1}{4}$ pound of fish equally among 3 cats.

That means she needs to divide $\frac{1}{4}$ into 3 equal parts. You can draw an area model to represent the problem.

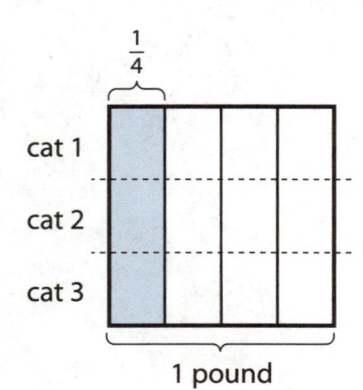

$\frac{1}{4} \div 3 = \frac{1}{12}$

If $\frac{1}{4}$ pound of fish is divided into 3 equal parts, each cat will receive $\frac{1}{3}$ of the $\frac{1}{4}$ pound of fish.

$\frac{1}{3} \times \frac{1}{4} = \frac{1}{12}$

✏️ **Circle** the multiplication equation that solves the division situation.

> **Think** What does dividing a whole number by a unit fraction mean?

Mr. Putnam wants to cut a 3-foot rope into $\frac{1}{4}$-foot sections.

To figure out how many sections he will get, Mr. Putnam thinks, "How many fourths are in 3?"

You can draw a number line to represent the 3 feet of rope. There are three 1-foot sections.

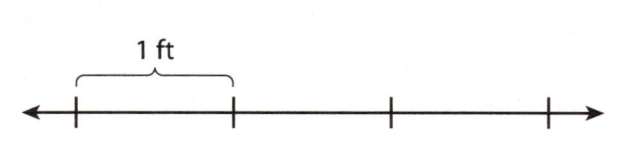

Look at the answer to this division problem. It is greater than 3, the number I started with!

You can mark fourths on the number line to represent $\frac{1}{4}$ foot. You can see there are twelve $\frac{1}{4}$-foot sections in 3 feet.

$3 \div \frac{1}{4} = 12$

You can also write a multiplication equation to show how many fourths are in 3. There are 4 fourths in each whole foot. To find the number of fourths in 3 feet, you can multiply.

$3 \times 4 = 12$

When you divide 3 by $\frac{1}{4}$, you are dividing 3 into parts smaller than 1. So there will be more than 3 of those parts.

▶ **Reflect**

1 Explain what it means to divide 5 by $\frac{1}{4}$.

_____

_____

Lesson 17 *Understand* Division with Unit Fractions **167**

Lesson 17 Guided Instruction

Think About: Using Unit Fractions in Division

Let's Explore the Idea Explore dividing a whole number by a unit fraction with the problem below.

Jemma made 5 cups of pancake batter. She uses a scoop measuring $\frac{1}{3}$ cup to pour batter onto the skillet to make large pancakes. How many pancakes can Jemma make?

The 5 rectangles below represent the 5 cups of pancake batter.

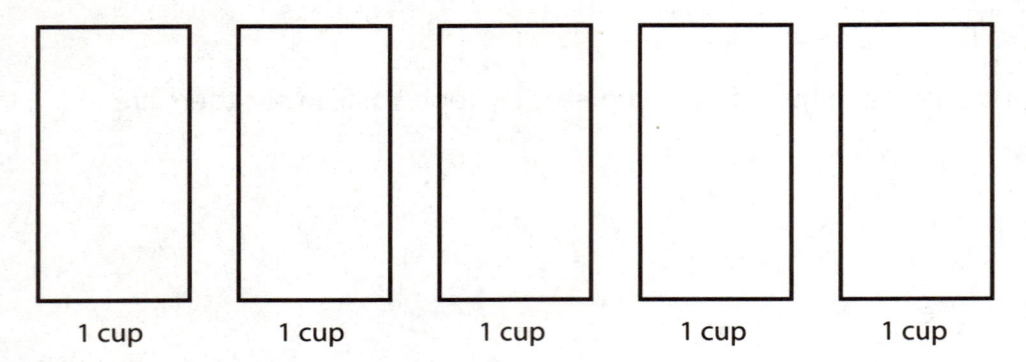

1 cup   1 cup   1 cup   1 cup   1 cup

2 You need to find out how many _____ there are in _____.

3 The scoop holds $\frac{1}{3}$ cup of batter. How many scoops are in 1 cup? _____

4 Divide each of the 5 rectangles into sections to show your answer to problem 3.

5 How many scoops are in 5 cups? _____

6 $5 \div \frac{1}{3} =$ _____

7 What multiplication equation will also solve this problem? _____

8 How is 5×3 related to $5 \div \frac{1}{3}$? _____

168 Lesson 17 Understand Division with Unit Fractions

Let's Talk About It

Solve the problems below as a group.

Suppose Jemma wanted to divide $\frac{1}{3}$ cup of pancake batter to make 4 mini pancakes. What fraction of a cup of batter will each pancake get?

9 The rectangle to the right shows 1 cup divided into 3 equal sections. How much does each section represent?

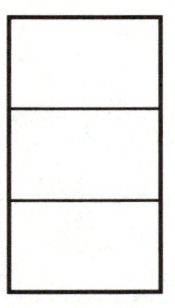

_____

10 Shade $\frac{1}{3}$ of the rectangle to show $\frac{1}{3}$ cup.

11 You need to divide $\frac{1}{3}$ cup equally to make 4 pancakes.

Divide each third of the rectangle vertically into 4 equal parts. Then shade $\frac{1}{4}$ of the rectangle to show 1 of the 4 pancakes.

12 The overlapping section shows the fraction of a cup of batter that each pancake will get. What is this fraction? _____

13 $\frac{1}{3} \div 4 =$ _____

14 What multiplication equation also solves $\frac{1}{4}$ of $\frac{1}{3}$? _____

15 How is $\frac{1}{3} \div 4$ related to $\frac{1}{3} \times \frac{1}{4}$? _____

_____

▶ **Try It Another Way** Explore dividing by a unit fraction using a common denominator.

Another way to think about dividing unit fractions is to write equivalent fractions with a common denominator. What is $5 \div \frac{1}{2}$?

16 Write 5 as a fraction with a denominator of 2. _____

17 Divide $\frac{10}{2}$ into equal groups of $\frac{1}{2}$. How many groups can you make? _____

18 $5 \div \frac{1}{2} =$ _____

Lesson 17 *Understand* Division with Unit Fractions **169**

Lesson 17 Guided Practice

Connect: Ideas About Dividing with Unit Fractions

Talk through these problems as a class. Then write your answers below.

19 Compare Draw a model to represent $\frac{1}{4} \div 4$ and a model to represent $\frac{1}{4} \times \frac{1}{4}$. Explain the relationship between the two expressions.

20 Analyze Helena said that $12 \div \frac{1}{3}$ is 4. Draw a model and use words to explain why Helena's statement is not reasonable.

21 Justify Show that $\frac{1}{2} \div 3 = \frac{1}{6}$ by using a model. Explain why the result is less than the number you started with, $\frac{1}{2}$.

Lesson 17 — Independent Practice

Apply: Ideas About Dividing with Unit Fractions

22 Put It Together Use what you have learned to complete this task.

Choose one of the following problems to solve. Circle the problem you choose.

> Drew wants to run at least 6 miles this month. He plans to run $\frac{1}{4}$ mile each day. How many days will it take Drew to run 6 miles?

> Maya made $\frac{1}{2}$ quart of strawberry jam. She plans to share it equally among 4 friends. How much jam will each friend get?

Part A Draw a model to represent the problem.

Part B Write a division equation and a multiplication equation that represent the problem.

_____

_____

_____

Lesson 17 *Understand* Division with Unit Fractions

Lesson 18 — Introduction
Divide Unit Fractions in Word Problems

NC.5.NF.7

Use What You Know

Now that you understand what it means to divide with unit fractions, take a look at this problem.

> Micah is running a 6-mile race. There are water stops every $\frac{1}{2}$ mile, including at the 6-mile finish line. How many water stops are there?

a. You want to find how many groups of $\frac{1}{2}$ there are in 6. Write a division expression that represents the problem. _____

b. You can use a number line to help you solve the problem. Finish labeling the number line below from 0 to 6.

0 ——————————————————— 6

c. Draw points on the number line to show the location of all the water stops.

d. How many $\frac{1}{2}$s are in 1? _____

e. How many $\frac{1}{2}$s are in 6? _____

f. Explain how you can use the number line to find the number of water stops there are. _____

172 Lesson 18 Divide Unit Fractions in Word Problems

▶▶ Find Out More

You can find how many half miles are in 6 miles in different ways.

- You can find $6 \div \frac{1}{2}$ by drawing a model.

  You can start by drawing 6 rectangles to represent the 6 miles.

  Start   Finish

  Then show the halves and count.

  Start   Finish

  first stop   last stop

  There are 12 halves, so there are 12 water stops.

- You can also find $6 \div \frac{1}{2}$ by writing a related multiplication equation.

  $6 \div \frac{1}{2} = ?$

  $? \times \frac{1}{2} = 6$

  $12 \times \frac{1}{2} = 6$

  So, $6 \div \frac{1}{2} = 12$.

▶ Reflect

1 Suppose the problem is changed so that there are water stops every $\frac{1}{4}$ mile, including the finish line. Explain why you would model this by dividing each rectangle into fourths instead of halves.

_____

_____

_____

_____

Lesson 18 Divide Unit Fractions in Word Problems

Lesson 18 Modeled and Guided Instruction

Learn About: Dividing a Fraction by a Whole Number

Read the problem below. Then explore different ways to understand dividing a unit fraction to solve word problems.

> Piper used $\frac{1}{5}$ meter of ribbon to create a border around a triangle. If each side of the triangle is the same length, how much ribbon did Piper use for each side?

▶ **Picture It** You can draw a picture to help understand the problem.

Draw a **1-meter** length of ribbon, then draw and label a $\frac{1}{5}$-**meter** length.

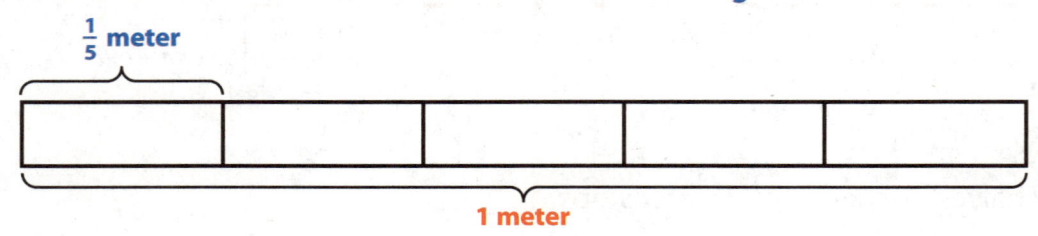

Divide the $\frac{1}{5}$-**meter** length into **3 equal parts** that represent the 3 sides of the triangle.

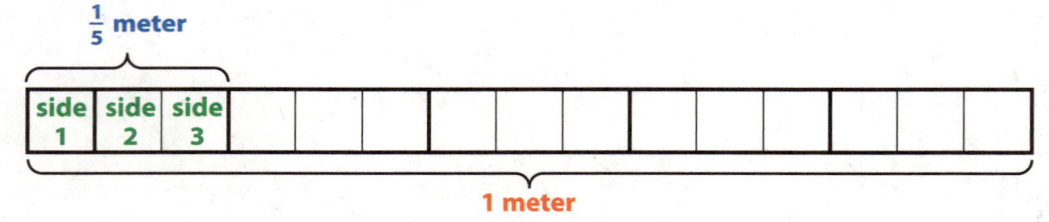

▶ **Model It** You can use a model to help understand the problem.

Draw and shade $\frac{1}{5}$ of a rectangle.

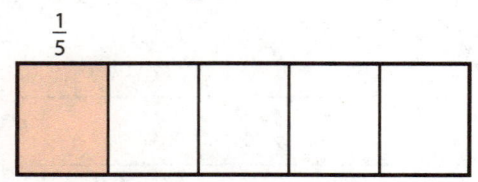

Divide the rectangle into 3 equal parts. Then shade one of the thirds of the shaded $\frac{1}{5}$.

174 Lesson 18 Divide Unit Fractions in Word Problems

Connect It Now you will solve the problem from the previous page using your understanding of unit fractions and the models.

2 Look at *Picture It* on the previous page. What information in the problem does the first diagram show?

_____

_____

3 Why does the second diagram show each bar divided into 3 equal parts?

_____

_____

4 When you divide $\frac{1}{5}$ meter into 3 equal parts, how long is each part? _____ meter
How do you know? _____

_____

5 How much ribbon did Piper use for each side of the triangle? _____

6 What division expression represents this problem? _____

7 What is $\frac{1}{5} \div 3$? _____

8 Describe how *Model It* on the previous page shows dividing a unit fraction by a whole number. _____

_____

_____

Try It Use what you just learned about dividing unit fractions to solve this problem. Show your work on a separate sheet of paper.

9 Tate has $\frac{1}{4}$ of a pizza. He wants to share the pizza equally with a friend. How much of the original whole pizza will each of them get? Draw a model and write an equation to represent and solve the problem.

_____

_____

Lesson 18 Divide Unit Fractions in Word Problems 175

Lesson 18 👥 **Modeled and Guided Instruction**

Learn About ▸ Dividing a Whole Number by a Fraction

Read the problem below. Then explore different ways to understand dividing a whole number by a unit fraction.

> Alex makes 2 pounds of bread dough. He separates the dough into $\frac{1}{4}$-pound loaves before baking them in the oven. How many loaves does he make?

▶ **Picture It** You can draw a model to help understand the problem.

Draw 2 rectangles to represent the 2 pounds of bread dough.

Show each pound divided into fourths.

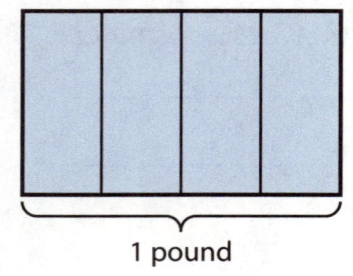

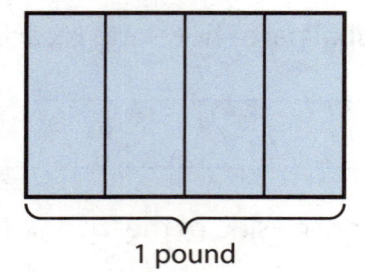

[Two rectangles, each divided into 4 equal vertical sections, labeled "1 pound" underneath each]

▶ **Model It** You can use a number line to help understand the problem.

Draw a number line and label it to show the 2 pounds of bread dough.

Mark the number line to divide each whole into fourths.

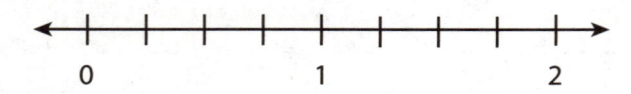

[Number line from 0 to 2 with tick marks at each fourth]

176 Lesson 18 Divide Unit Fractions in Word Problems ©Curriculum Associates, LLC Copying is not permitted.

Connect It Now you will solve the problem from the previous page using the models.

10 Write a division expression that represents the problem. _____

11 How many fourths are in one whole? _____

How many fourths are in two wholes? _____

12 Use the information in problem 11 to solve the problem from the previous page.

Using words: There are _____ fourths in 2.

Using numbers: $2 \div \frac{1}{4} =$ _____

Alex makes _____ loaves of bread.

13 What multiplication equation could you write to check your answer? _____

14 Describe a model you can use to divide a whole number by a unit fraction.

_____

_____

_____

Try It Use what you just learned about dividing whole numbers by unit fractions to solve this problem. Show your work on a separate sheet of paper.

15 Stacy has 4 sheets of paper to make cards. Each card requires $\frac{1}{2}$ sheet of paper. How many cards can Stacy make? Choose a model to solve the problem. Then explain why you chose that model.

Lesson 18 Divide Unit Fractions in Word Problems

Lesson 18 Guided Practice

Practice: Dividing Unit Fractions in Word Problems

Study the example below. Then solve problems 16–18.

Example

Sierra has a photo album with 3 empty pages. Each photo uses $\frac{1}{6}$ of an album page. How many photos can Sierra put on the empty pages?

Look at how you could show your work using rectangles.

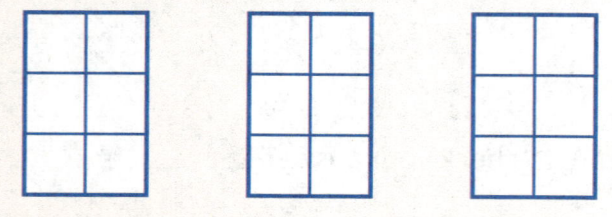

6 photos will fit on one page, so 18 photos will fit on 3 pages.

Solution 18 photos

> The student used a model to visualize the problem.

> **Pair/Share**
> What related equations can you write to represent the problem?

16 Corrine picked $\frac{1}{4}$ gallon of blackberries. She poured the berries equally into 4 containers. What fraction of a gallon is in each container?

Show your work.

> Can I draw a model to help understand the problem?

> **Pair/Share**
> How will the answer compare to $\frac{1}{4}$ gallon?

Solution _____

Lesson 18 Divide Unit Fractions in Word Problems

17 Cooper's USB drive is $\frac{1}{2}$ full with 5 video files. Each video file is the same size. What fraction of the USB drive does 1 video file use?

Show your work.

How could I represent this problem using an equation?

Pair/Share
How can you check your answer?

Solution _____

18 Devonte is studying for a history test. He uses $\frac{1}{8}$ of a side of one sheet of paper to write notes for each history event. He fills 2 full sides of one sheet of paper. Which expression could be used to find how many events Devonte makes notes for? Circle the letter of the correct answer.

A $2 \times \frac{1}{8}$

B $2 \div \frac{1}{8}$

C $\frac{1}{8} \times 2$

D $\frac{1}{8} \div 2$

Is this problem like any problem I've seen before?

Barry chose **D** as the correct answer. How did he get that answer?

_____
_____
_____
_____

Pair/Share
Does Barry's answer make sense?

©Curriculum Associates, LLC   Copying is not permitted.   Lesson 18 Divide Unit Fractions in Word Problems   **179**

Lesson 18 — Independent Practice

Practice: Dividing Unit Fractions in Word Problems

Solve the problems.

1 Elise picks 6 pounds of apples. She uses $\frac{1}{2}$ pound of apples to make 1 container of applesauce. How many containers of applesauce can Elise make with all the apples?

    A 12 containers

    B $6\frac{1}{2}$ containers

    C $5\frac{1}{2}$ containers

    D 3 containers

2 Students are running in a relay race. Each team will run a total of 3 miles. Each member of a team will run $\frac{1}{3}$ mile.

How many students will a team need to complete the race? Circle the correct number below.

    $\frac{1}{9}$ 3   9   12   36

You may use the number line to help find your answer.

3 Mr. Bernstein will cut 8 pies into pieces that are each $\frac{1}{6}$ of the whole. After he cuts the 8 pies, how many pieces will Mr. Bernstein have? _____ pieces

4 Marina has a pattern to make bows that requires $\frac{1}{4}$ yard of ribbon for each bow.

Part A Fill in the table to show how many bows she can make from a given length of ribbon.

Ribbon Length (yards)	Number of Bows
1	
2	
3	
4	

Part B Use words or an equation to describe a rule to find the number of bows Marina can make if you know how many yards of ribbon she has.

_____

_____

Part C Use your rule to find how many bows Marina can make if she has 18 yards of ribbon.

Answer _____ bows

✓ **Self Check** Go back and see what you can check off on the Self Check on page 103.

Lesson 18 Divide Unit Fractions in Word Problems **181**

Unit 2 MATH IN ACTION

Introduction
Use Fractions

SMP1 Make sense of problems and persevere in solving them.

Study an Example Problem and Solution

In this lesson, you will use what you know about fractions to solve real-world problems. Look at this problem and one solution.

Solar Lights

G.O. and his neighbors are taking steps to save energy and water. G.O.'s street is $1\frac{1}{2}$ miles long. They plan to install solar lights along the sidewalk. Read G.O.'s plan.

Solar Light Plan
- Install the lights at equal intervals.
- Use a fraction of the whole length to choose the interval length.
- Use a fraction greater than $\frac{1}{8}$ but less than $\frac{1}{2}$.

Choose an appropriate fraction. Find the length of the interval. Tell how many solar lights are needed and where along the street the lights should be located.

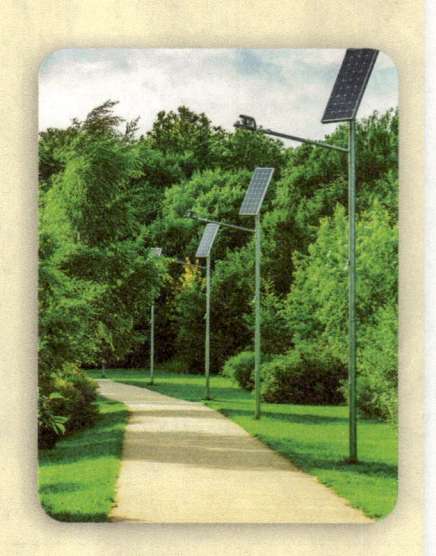

Read the sample solution on the next page. Then look at the checklist below. Find and mark parts of the solution that match the checklist.

Problem-Solving Checklist

☐ Tell what is known.
☐ Tell what the problem is asking.
☐ Show all your work.
☐ Show that the solution works.

a. **Circle** something that is known.
b. **Underline** something that you need to find.
c. **Draw a box around** what you do to solve the problem.
d. **Put a checkmark** next to the part that shows the solution works.

G.O.'s Solution

Hi, I'm G.O. Here's how I solved this problem.

▷ **I know the length of the street.** I have to use the length and a fraction between $\frac{1}{8}$ and $\frac{1}{2}$ to find the distance between the lights.

▷ **I can use $\frac{1}{4}$.** The fractions $\frac{1}{8}$, $\frac{1}{4}$, and $\frac{1}{2}$ all have the same numerator, so I can look at the denominators to compare.

Since 4 is < 8, $\frac{1}{4} > \frac{1}{8}$.

Since 4 is > 2, $\frac{1}{4} < \frac{1}{2}$.

▷ **I know that** the distance is a fraction of the whole length, so I multiply $\frac{1}{4}$ by $1\frac{1}{2}$.

$1\frac{1}{2} = \frac{3}{2}$ and $\frac{1}{4} \times \frac{3}{2} = \frac{3}{8}$

The interval length is $\frac{3}{8}$ mile.

"Fraction of" means multiply by a fraction.

▷ **Now I can** make a number line to find where the lights will go and how many are needed. It represents the street, so it goes to $1\frac{1}{2}$.

I divided the number line into eighths so I can count by $\frac{3}{8}$.

0    $\frac{3}{8}$ $\frac{6}{8}$ 1 $\frac{9}{8}$ $\frac{12}{8}$

▷ **I put a light at 0**, which is one end of the street. Then I marked each $\frac{3}{8}$ mile along the number line.

▷ **The number line shows** that there will be 5 lights installed at intervals of $\frac{3}{8}$ mile. The locations are at:

0 miles, $\frac{3}{8}$ mile, $\frac{6}{8}$ mile, $\frac{9}{8}$ ($1\frac{1}{8}$ miles), and $\frac{12}{8}$ ($1\frac{1}{2}$ miles).

▷ **There are 5 lights but only 4 intervals.** Each interval is $\frac{3}{8}$ mile long and $4 \times \frac{3}{8} = \frac{12}{8}$. That's the same as $1\frac{1}{2}$ miles.

My answer makes sense because it fits the information in the problem.

Unit 2 Math in Action — Modeled and Guided Instruction

Try Another Approach

There are many ways to solve problems. Think about how you might solve the Solar Lights problem in a different way.

Solar Lights

G.O. and his neighbors are taking steps to save energy and water. G.O.'s street is $1\frac{1}{2}$ miles long. They plan to install solar lights along the sidewalk. Read G.O.'s plan.

Solar Light Plan
- Install the lights at equal intervals.
- Use a fraction of the whole length to choose the interval length.
- Use a fraction greater than $\frac{1}{8}$ but less than $\frac{1}{2}$.

Choose an appropriate fraction. Find the length of the interval. Tell how many solar lights are needed and where along the street the lights should be located.

▶ **Plan It** Answer these questions to help you start thinking about a plan.

A. What are some other fractions that you can use?

_____

_____

B. What can you do if you want to use more lights? Fewer lights?

_____

_____

184 Unit 2 Math in Action Use Fractions

▶ **Solve It** Find a different solution for the Solar Lights problem. Show all your work on a separate sheet of paper.

You may want to use the problem-solving tips to get started.

Problem-Solving Tips

- **Models**

  0 — 1 — 1½

- **Word Bank**

  fraction   numerator   multiply
  interval   denominator product

- **Sentence Starters**

  • I can multiply _____

  • Each interval _____

Problem-Solving Checklist

Make sure that you . . .
☐ tell what you know.
☐ tell what you need to do.
☐ show all your work.
☐ show that the solution works.

▶ **Reflect**

Use Mathematical Practices As you work through the problem, discuss these questions with a partner.

- **Persevere** What is your first step? What will you do next?

- **Repeated Reasoning** How can you use what you know about the denominators of unit fractions to find an appropriate fraction?

Unit 2 Math in Action — Guided Practice

Discuss: Models and Strategies

Read the problem. Write a solution on a separate sheet of paper. Remember, there can be lots of ways to solve a problem!

Plant Shrubs

The neighborhood has a small piece of common land that is now covered with grass. To save water, the neighbors will plant shrubs on part of the common area. Read G.O.'s notes.

Planting Notes
- Plant shrubs on a rectangular area a little more than half of the common land area.
- One side of the shrub section has a length greater than 8 feet and less than 9 feet.

Common Land: 10 feet by 10 feet

What is the area of the part where G.O. and his neighbors will plant shrubs?

▶ Plan It and Solve It Find a solution to the Plant Shrubs problem.

Find the length, width, and area of the part of the common land that will be used to plant shrubs.

- Find the area of the common land.
- Find a length and width of a rectangle that will create an area that is a little more than half the area of the common land.

Problem-Solving Tips

- **Questions**
  - What are some fractions equivalent to $\frac{1}{2}$?
  - What are some fractions that are a little more than $\frac{1}{2}$?

- **Word Bank**

area	multiply	numerator
fraction	product	denominator

Problem-Solving Checklist
Make sure that you . . .
☐ tell what you know.
☐ tell what you need to do.
☐ show all your work.
☐ show that the solution works.

▶ Reflect

Use Mathematical Practices As you work through the problem, discuss these questions with a partner.

- **Reason Mathematically** How can you compare fractions to find a fraction a little more than $\frac{1}{2}$?
- **Use Models** What models can you use to help you visualize the problem?

Unit 2 Math in Action — Independent Practice

Persevere ▶ On Your Own

Read the problems. Write a solution on a separate sheet of paper. Remember, there are many different ways to solve a problem.

Water Shrubs

G.O. and his neighbors clear an area $8\frac{1}{2}$ feet by $6\frac{1}{4}$ feet to plant the shrubs. Now they have to decide how many shrubs to plant and how much water to use on the shrubs. Read G.O.'s planting instructions.

Shrub Planting Instructions
- Each shrub needs an area of about 2 square feet.
- Each shrub will need about $1\frac{1}{4}$ gallons of water a week.

How many shrubs should G.O. plant?

How much water will the shrubs need?

▶ **Solve It** Help G.O. make a plan for planting shrubs.
- Tell how many shrubs G.O. should plant and why you chose this number.
- Find the amount of water this number of shrubs will need in a week.

▶ **Reflect**

Use Mathematical Practices After you complete the task, choose one of these questions to discuss with a partner.

- **Reason Mathematically** How did you decide the number of shrubs that G.O. should plant?

- **Make an Argument** How could you justify the number of shrubs that you suggested?

Use Compost

A local nursery hears about the shrub planting project that G.O. and his neighbors are planning. The nursery gives them 50 pounds of compost to use. G.O. reads about using compost on a website.

About how many shrubs can G.O. plant with the compost that the nursery gave him?

Using Compost to Plant

When you plant a shrub, it can help to mix the soil with some compost. You can use a scoop of compost for each shrub. An average scoop of compost is between $\frac{1}{4}$ and $\frac{1}{2}$ pound.

▶ **Solve It** Help G.O. estimate how many shrubs he can plant with the compost.
- Decide on a fraction of a pound to use as the weight of an average scoop.
- Show how to use this fraction to find the number of shrubs that can be planted. Explain.

▶ **Reflect**
Use Mathematical Practices After you complete the task, choose one of these questions to discuss with a partner.
- **Make Sense of Problems** How did you use each of the numbers given in the problem?
- **Persevere** Why might you try using different fractions before giving your final answer? Explain.

Unit 2 Assessment
Interim Assessment

Solve the problems.

1. What is the area of the rectangle shown below?

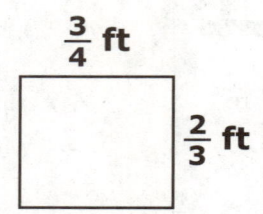

   $\frac{3}{4}$ ft

   $\frac{2}{3}$ ft

   A $\frac{12}{6}$ square feet

   B $\frac{5}{7}$ square foot

   C $\frac{8}{9}$ square foot

   D $\frac{6}{12}$ square foot

2. Look at the equations below. Which equation is true?

   A $3\frac{4}{5} + 6\frac{2}{10} = 10$

   B $22\frac{7}{8} - 16\frac{1}{4} = 6\frac{6}{8}$

   C $\frac{2}{3} \times \frac{1}{6} = \frac{4}{6}$

   D $6 \div \frac{1}{5} = \frac{6}{5}$

3. Liam says that $n \times \frac{3}{7}$ is greater than $\frac{3}{7}$ because multiplication always increases a number. Which of the following values for n prove that Leo is incorrect?

   A $1\frac{1}{8}$

   B $\frac{2}{3}$

   C $\frac{9}{4}$

   D $\frac{5}{5}$

4. Look at each fraction. Which fraction can be placed in the box to make a true inequality?

   $\frac{4}{11} \times \boxed{} < \frac{4}{11}$

   A $\frac{5}{3}$

   B $\frac{3}{3}$

   C $\frac{1}{3}$

   D $\frac{4}{3}$

5. How many pounds of peanuts will each person get if 3 people share $\frac{1}{5}$ pound of peanuts equally?

Which pair of division and multiplication expressions can be used to represent the problem?

A $\frac{1}{5} \div 3$ and $3 \times \frac{1}{15}$

B $3 \div \frac{1}{5}$ and $\frac{1}{5} \times 3$

C $\frac{1}{5} \div 3$ and $\frac{1}{15} \times \frac{1}{5}$

D $3 \div \frac{5}{1}$ and $\frac{1}{15} \times \frac{5}{1}$

6. José painted a square-shaped mural. The length of each side of the canvas is $1\frac{1}{3}$ yards.

What is the area of the mural in square yards?

Unit 2 Interim Assessment continued

Performance Task

Answer the questions and show all your work on separate paper.

The Drama Club is painting sets for their next play. Blue and green paint is on sale at the hardware store, so the students have sketched the design for a set that will use only those two paint colors. How much of each color will they need to buy in order to paint the set?

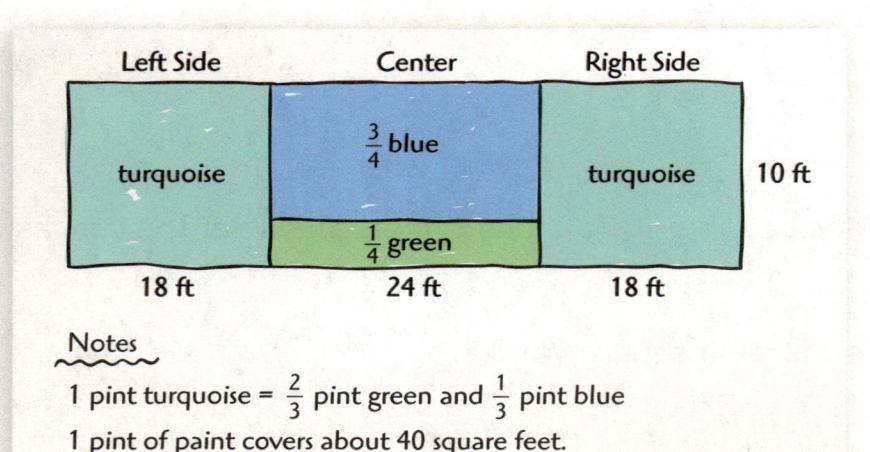

Checklist

Did you . . .
- [ ] use the information in the diagram and the notes?
- [ ] organize your information?
- [ ] check the accuracy of your solution?

Left Side: turquoise, 18 ft
Center: $\frac{3}{4}$ blue, $\frac{1}{4}$ green, 24 ft
Right Side: turquoise, 18 ft
Height: 10 ft

Notes
1 pint turquoise = $\frac{2}{3}$ pint green and $\frac{1}{3}$ pint blue
1 pint of paint covers about 40 square feet.

▶ **Reflect**

Use Mathematical Practices After you complete the task, choose one of the following questions to answer.

- **Persevere** What was your first step in solving this problem?
- **Model** How did you use fractions to help you solve this problem?

Unit 3
Operations and Algebraic Thinking

Let's learn about expressions and relationships between numbers.

Real-World Connection Sometimes you can solve a problem by completing the steps in any order. Other times you need to complete the steps in a certain order. For example, consider making a pizza with a vegetable mixture and pepperoni. You could first put the vegetable mixture on the pizza, and then top with pepperoni. Or you could first put on the pepperoni and then top with the vegetable mixture. Either way, the pizza will be tasty! Suppose you are given the math problem "double 8 + 12, then add 43." The order in which you complete the steps is important. The expressions $2 \times 8 + 12 + 43$ and $2 \times (8 + 12) + 43$ have different values. You must follow the steps in a certain order to find the correct value.

In This Unit You will write numerical expressions and find their value using the order of operations and properties of real numbers. You will also identify and create number patterns, describe the relationships between numbers in the patterns, and plot the numbers on a graph.

✓ Self Check

Before starting this unit, check off the skills you know below. As you complete each lesson, see how many more skills you can check off!

I can:	Before this unit	After this unit
evaluate expressions, for example: $48 \div (6 + 10)$ has a value of 3.	☐	☐
write expressions, for example: "subtract 5 from 12, then multiply by 4" can be written as $(12 - 5) \times 4$.	☐	☐
determine if two expressions are the same, for example: $14 + 9 + 7$ is the same as $7 + 14 + 9$.	☐	☐

find the relationship between two number patterns, for example:
    Pattern A: 0, 2, 4, 6, 8, . . .
    Pattern B: 0, 8, 16, 24, 32, . . .
Each term in Pattern B is 4 times the corresponding term in Pattern A.
	☐	☐

list ordered pairs for two number patterns and graph the relationship on the coordinate plane, for example: ordered pairs for Patterns A and B above are (0, 0), (2, 8), (4, 16), (6, 24), (8, 32).	☐	☐

Lesson 19 — Introduction
Evaluate and Write Expressions

NC.5.OA.2

Use What You Know

You know about the order of operations. Now you will see how using parentheses in an expression can change the value of the expression. Take a look at this problem.

> Maria and her friend go to a movie. At the snack stand, they each get a drink that costs $5 and a popcorn that costs $8. Maria pays for her friend. How much does Maria pay altogether?

a. What operation do you use to find the cost of a drink and a popcorn for one person?
_____

b. Write an expression for the cost of a drink and popcorn for one person. _____

c. How does the cost for two people compare to the cost for one person?
_____

d. Explain how you can find the cost for two people. What do you need to do first?
_____
_____

e. How much does Maria pay for a drink and popcorn for two people? _____

f. Maria thought she could use the equation $2 \times 5 + 8 = 18$ to find the cost. Explain why she is not correct.
_____
_____
_____

▶▶ Find Out More

To **evaluate** an expression means to find its value. To evaluate the expression $2 \times 5 + 8$, you first multiply 2×5, then add 8. But what if you wanted to add 5 and 8 and then multiply by 2? You could use **parentheses** in the expression to tell which operation to do first.

Parentheses are a type of grouping symbol. Grouping symbols tell which operation to do first. Fraction bars are another type of grouping symbol.

Parentheses	$2 \times (5 + 8)$	First, add $5 + 8$ because it is inside the parentheses. Then multiply by 2.
Fraction Bar	$\dfrac{2 + 8}{5}$	The fraction bar groups the numerator separately from the denominator. First evaluate the numerator. Then divide.

Look at the problem on the previous page. Maria wants to first find the cost of a drink and popcorn for 1 person, then double it to find the cost for 2 people. Maria could write the expression $2 \times (5 + 8)$ or $(5 + 8) \times 2$.

One way to read $2 \times (5 + 8)$ is "2 times the sum of 5 and 8." Another way is "twice the sum of 5 and 8." You can also think of $2 \times (5 + 8)$ as "adding 5 and 8, then multiplying by 2."

Adding parentheses into an expression can change its value. The expressions $2 \times 5 + 8$ and $2 \times (5 + 8)$ do not have the same value.

$2 \times 5 + 8$ $2 \times (5 + 8)$
$10 + 8$ 2×13
18 26

▶ Reflect

1 What should you look for to indicate a group? How do you evaluate an expression if you see a grouping symbol?

_____

_____

_____

_____

Lesson 19　Modeled and Guided Instruction

Learn About: Evaluating Expressions

Read the problem below. Then explore how to evaluate expressions that use grouping symbols.

> There were 24 students on a field trip to the aquarium. There were also 8 adults on the trip. The expression 0.50 × (24 + 8) represents the cost in dollars to buy everyone a 50-cent souvenir eraser. What is the total cost of the erasers?

▶ **Picture It** You can use a picture to help understand the problem.

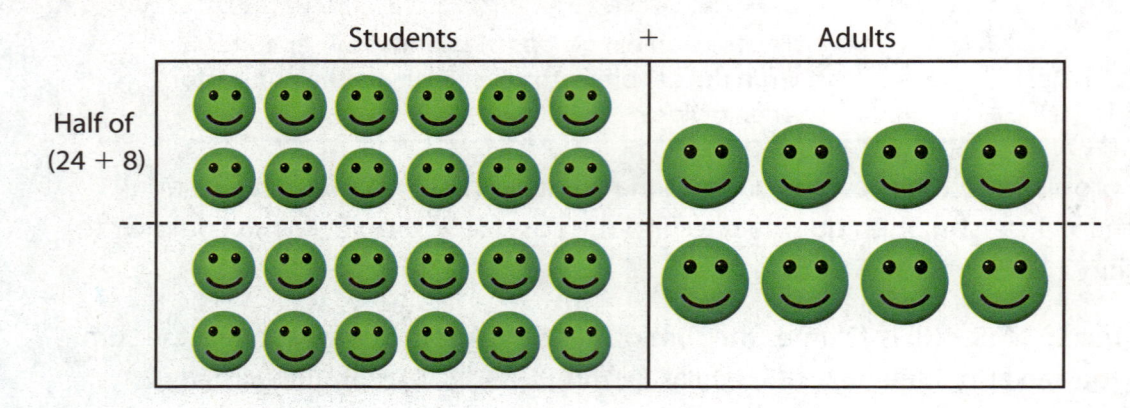

Students　+　Adults

Half of (24 + 8)

▶ **Model It** You can use words to help understand the problem.

0.50　　×　　(24 + 8)
↑　　　↑　　　↑
Half　　of　　the sum of the number of students and the number of adults

196　Lesson 19 Evaluate and Write Expressions

©Curriculum Associates, LLC　Copying is not permitted.

▶ **Connect It** Now you will solve the problem from the previous page using the picture and words.

2 Describe one way you could read the expression 0.50 × (24 + 8).

_____

3 How could you use *Picture It* on the previous page to evaluate 0.50 × (24 + 8)?

_____

_____

4 Evaluate 0.50 × (24 + 8) to find the cost in dollars of the erasers. _____

5 Morgan sees a different way to evaluate 0.50 × (24 + 8). She instead writes the expression $\frac{24 + 8}{2}$. Why does her method work?

_____

_____

6 The expression 3 × (number of students + number of adults) represents the cost in dollars for another group to go to the dolphin show at the aquarium. Describe how the cost compares to the total number of students and adults.

_____

_____

▶ **Try It** Use what you just learned about evaluating expressions to solve these problems. Show your work on a separate sheet of paper.

7 Describe what happens when you multiply a sum by 2.

_____

8 Sara buys a shirt that regularly costs $12 and a pair of pants that regularly costs $26. They are on sale, so she only needs to pay half the regular cost. Evaluate the expression $\frac{1}{2}$ × (12 + 26) to find Sara's cost in dollars. _____

Lesson 19 Evaluate and Write Expressions

Lesson 19 Modeled and Guided Instruction

Learn About > Writing Expressions

Read the problem below. Then explore how to write numerical expressions.

> Write a numerical expression to represent the following phrase.
>
> *15 minus the sum of 6 and 7*

▶ **Picture It** You can use a picture to help understand the problem.

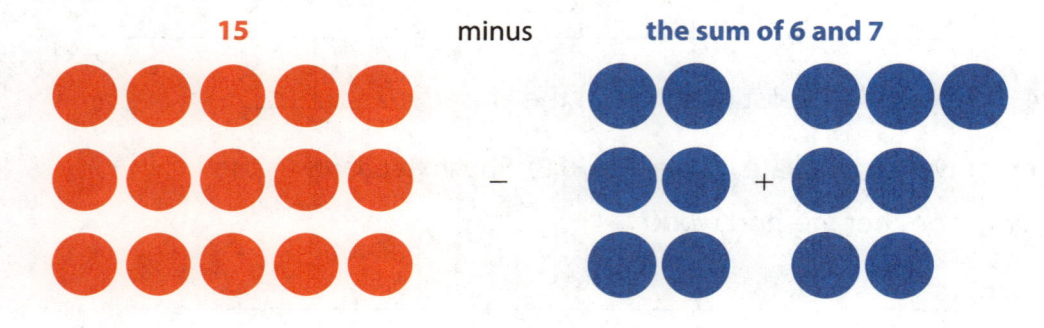

15   minus   the sum of 6 and 7

▶ **Model It** You can think about what the words mean to help understand the problem.

15   minus   the sum of 6 and 7
 ↓   ↓
Minus means   A sum is the
to subtract.   result of
               addition. So
               add 6 and 7.

198 Lesson 19 Evaluate and Write Expressions

Connect It Now you will solve the problem from the previous page using the picture and words.

9 In the expression "15 minus the sum of 6 and 7," do you add or subtract first? Why?

_____

_____

10 When you write a numerical expression, how can you show what operation to do first?

_____

_____

11 Write a numerical expression for "15 minus the sum of 6 and 7." _____

12 Harper wrote the expression 15 − 6 + 7 to represent "15 minus the sum of 6 and 7." Evaluate 15 − 6 + 7 and then explain why Harper's expression is incorrect.

_____

_____

_____

13 Omar wrote 3 + (4 × 6) to represent the phrase "3 more than the product of 4 and 6." Did Omar need to use a grouping symbol? Explain.

_____

_____

_____

Try It Use what you just learned about writing numerical expressions to solve these problems. Show your work on a separate sheet of paper.

14 Write a numerical expression to represent "2 times the difference of 8 and 1." Then evaluate your expression. _____

15 Write a numerical expression to represent "15 divided by the sum of 1 and 4." Then evaluate your expression.

_____

Lesson 19 Evaluate and Write Expressions

Lesson 19 Guided Practice

Practice: Writing and Evaluating Expressions

Study the example below. Then solve problems 16–18.

Example

Insert parentheses to make the following equation true.

$15 - 7 - 2 = 10$

Look at how you could show your work.

$(15 - 7) - 2 = 8 - 2 = 6$

$6 \neq 10$

$15 - (7 - 2) = 15 - 5 = 10$

$10 = 10$

Solution $15 - (7 - 2) = 10$

The student used trial and error to answer the question.

Pair/Share
How many different ways can you group the numbers?

---

16 Carol sells bracelets and pairs of earrings at a craft fair. Each item sells for $8. Write and evaluate an expression to show how much money Carol will make if she sells 23 bracelets and 17 pairs of earrings.

Show your work.

How many items will Carol sell altogether?

Pair/Share
What other ways could you solve the problem?

Solution _____

200 Lesson 19 Evaluate and Write Expressions

17 Write numerical expressions for "the product of 3 and 2, plus 5" and "3 times the sum of 2 and 5." Which expression has a greater value?

Show your work.

The comma is a clue to where to put the grouping symbol.

Pair/Share
When do you use parentheses in an expression?

Solution _____

18 Which expression represents "the quotient of 10 and 2, plus 3"? Circle the letter of the correct answer.

A $10 \div (2 + 3)$

B $\dfrac{10}{2 + 3}$

C $(10 \times 2) + 3$

D $\dfrac{10}{2} + 3$

Jason chose **A** as the correct answer. How did he get that answer?

_____
_____
_____
_____

What does the word "quotient" mean?

Pair/Share
Is Jason's answer reasonable?

Lesson 19 Independent Practice

Practice Writing and Evaluating Expressions

Solve the problems.

1 Kris ran 3 miles each day for 7 days in a row. One day, she ran an extra $\frac{1}{2}$ mile. Which expression represents how many miles Kris ran altogether?

A $3 + 7 + \frac{1}{2}$

B $3 \times 7 + \frac{1}{2}$

C $3 \times 7 + 3\frac{1}{2}$

D $\left(3 + \frac{1}{2}\right) \times 7$

2 Which expression does NOT represent the statement "half the difference of 20 and 8"?

A $\frac{20 - 8}{2}$

B $(20 - 8) \div 2$

C $(20 - 8) \times \frac{1}{2}$

D $20 - 8 \div 2$

3 Which expression has a value of 6? Circle the letter for all that apply.

A $4 + 8 \div 2$

B $(4 + 8) \div 2$

C $4 + (8 \div 2)$

D $(8 + 4) \div 2$

E $8 + 4 \div 2$

4 Adam is 2 years old. His sister Lina is 1 year less than three times his age. Write a numerical expression for Lina's age. _____

5 Several expressions are shown below. Decide if the value of the expression is less than, equal to, or greater than 18. Write each expression in the correct category in the chart.

$\frac{1}{3} \times (9 \times 2)$ $(9 \times 2) \times 1$ $(9 \times 2) \div 3$ $22 - (9 \times 2)$

$(9 \times 2) + 7$ $(9 \times 2) \times 4$ $1 \times 9 \times 2$ $3 \times (9 \times 2)$

Less than 18	Equal to 18	Greater than 18

6 Compare the expressions $8 \times 3 + 4$ and $8 \times (3 + 4)$. Explain how to evaluate each expression. Then tell which expression has the greater value.
Show your work.

✓ **Self Check** Go back and see what you can check off on the Self Check on page 193.

Lesson 20 · Introduction
Evaluate Expressions Using Properties

NC.5.OA.2

Use What You Know

You know about evaluating expressions that use grouping symbols. Now you will use number properties to evaluate expressions. Look at this problem.

> A bread recipe calls for 4 cups of white flour and 7 cups of wheat flour for each batch. How many cups of flour are needed to make three batches?

a. What operation do you use to find the number of cups of flour in one batch of bread?
_____

b. Write an expression for the number of cups of flour in one batch of bread. _____

c. Compare the number of cups of flour for three batches to the number of cups of flour for one batch.
_____

d. Explain how you can use a grouping symbol to write an expression for the number of cups of flour in three batches of bread.
_____
_____

e. How much flour does the baker need to make three batches of bread? _____

f. Explain why the expression 4 + 4 + 4 + 7 + 7 + 7 correctly finds the number of cups of flour needed to make three batches of bread.
_____
_____

204 Lesson 20 Evaluate Expressions Using Properties

Find Out More

The expression 3 × (4 + 7) represents the number of cups of flour needed for three batches of bread. One way to evaluate 3 × (4 + 7) is to add 4 and 7, then multiply by 3. Another way is to use properties.

You can use the **distributive property** to simplify the expression. You know that when one factor of a product is written as a sum, you can multiply each addend by the other factor. So, 3 × (4 + 7) = (3 × 4) + (3 × 7).

The distributive property is a number property. The table lists several number properties.

Associative Property	Changing the grouping of three or more addends does not change the sum. Changing the grouping of three or more factors does not change the product.	4 + (3 + 2) = (4 + 3) + 2 3 × (2 × 4) = (3 × 2) × 4
Commutative Property	Changing the order of the addends does not change the sum. Changing the order of the factors does not change the product.	2 + 4 = 4 + 2 3 × 4 = 4 × 3
Distributive Property	When one of the factors of a product is written as a sum, multiplying each addend by the other factor before adding does not change the product.	2 × (3 + 4) = (2 × 3) + (2 × 4)

Look at the problem on the previous page. After finding the number of cups of flour in 1 batch of bread, you can triple that amount to find the number of cups of flour in 3 batches. You can write the expression 3 × (4 + 7).

You can evaluate the expression by adding 4 and 7 and then multiplying by 3. Or you can multiply each addend by the factor 3 and then add the products. The expressions 3 × (4 + 7) and (3 × 4) + (3 × 7) have the same value.

3 × (4 + 7)   (3 × 4) + (3 × 7)
3 × 11   12 + 21
33   33

▶ Reflect

1 How do you evaluate an expression using the distributive property? When might using the distributive property make an expression easier to evaluate?

_____

_____

_____

Lesson 20 Evaluate Expressions Using Properties

Lesson 20 **Modeled and Guided Instruction**

Learn About: Evaluating Expressions with the Distributive Property

Read the problem below. Then explore how to evaluate expressions using the distributive property.

> There are 48 visitors and 8 tour guides at the natural history museum. The museum has strict rules about the total number of people allowed in the butterfly exhibit at one time. The expression $\frac{1}{4} \times (48 + 8)$ represents the number of people that can enter the exhibit at once. What is the total number of people that can enter the exhibit at one time? If the number of visitors and tour guides is the same in each group, how many are visitors and how many are tour guides?

▶ Picture It   You can use a picture to help understand the problem.

$\frac{1}{4}$ of 48 and 8

▶ Model It   You can use equations to help understand the problem.

$\frac{1}{4}$ × (48 + 8) = $\left(\frac{1}{4} \times 48\right)$ + $\left(\frac{1}{4} \times 8\right)$

| One fourth | of | the number of visitors and the number of tour guides | is equal to | one fourth of the number of visitors | and | one fourth of the number of tour guides |

206 Lesson 20 Evaluate Expressions Using Properties

©Curriculum Associates, LLC   Copying is not permitted.

Connect It Now you will solve the problem on the previous page.

2 Write one way to read the expression $\frac{1}{4} \times (48 + 8)$.

_____

3 Explain how you could use *Picture It* to evaluate $\frac{1}{4} \times (48 + 8)$.

_____

_____

_____

4 Using the distributive property, what is another way to write the expression? Evaluate the expression to solve the problem.

_____

_____

5 Jesse uses the commutative property to write the expression $\frac{1}{4} \times (8 + 48)$. Does Jesse's expression represent the problem? Explain.

_____

_____

6 John uses the associative property to write the expression $\left(\frac{1}{4} \times 48\right) + 8$. Does John's expression represent the problem? Explain.

_____

_____

Try It Use what you learned about the distributive property to solve these problems.

7 Olivia buys two books for half their regular prices. One book regularly costs $14, and the other regularly costs $18. Evaluate the expression $\frac{1}{2} \times (14 + 18)$ to find the cost in dollars. _____

8 When one factor of an expression is written as a difference, can you use the distributive property to evaluate the expression? Evaluate $\frac{1}{3} \times (99 - 30)$.

_____

_____

Lesson 20 Evaluate Expressions Using Properties **207**

Lesson 20 Modeled and Guided Instruction

Learn About: Evaluating Expressions with Number Properties

Read the problem below. Then explore how and when to use number properties to evaluate expressions.

> Deon has $20. He spends $3 on bus fare and $9 on an entry ticket to the aquarium. The expression 20 − (3 + 9) represents how much money he has left after he buys both items. What is the amount of money that Deon has left?

▶ **Picture It** You can use a picture to help understand the problem.

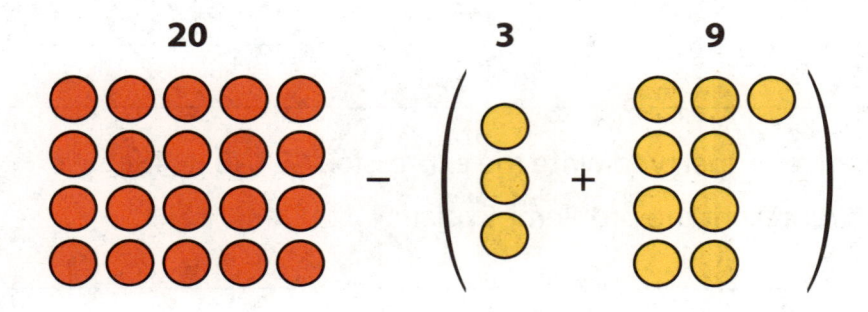

20 − (3 + 9)

▶ **Model It** You can think about what the expression means in words to help understand the problem.

20   −   (3 + 9)
↑   ↑   ↑

Deon starts with $20.   Subtract because Deon spends money.   Deon spends $3 on bus fare and $9 for a ticket.

208   Lesson 20 Evaluate Expressions Using Properties

Connect It Now you will solve the problem on the previous page.

9 What are two different ways to read the expression 20 − (3 + 9)?

_____

10 Explain how to evaluate 20 − (3 + 9) to find how much money Deon has left.

_____

_____

11 Evaluate 20 − (3 + 9) to find how much money Deon has left. _____

12 Jenna uses the associative property. She writes and evaluates the expression (20 − 3) + 9. Does Jenna's method work? Explain.

_____

_____

13 Could you use the commutative property when you evaluate the expression? Explain.

_____

_____

14 For which of the four operations do the associative and commutative properties work?

_____

Try It Use what you just learned about evaluating expressions with number properties to solve these problems. Tell which properties you use and why.

15 Carly earns $25 each week. She saves half the money she earns. The expression $4 \times \left(25 \times \frac{1}{2}\right)$ represents how much money she has saved after 4 weeks. How much money does Carly have saved after 4 weeks? Explain.

_____

_____

16 Evaluate the expression (13 + 49) + 7. Explain how you found your answer.

_____

_____

Lesson 20 Evaluate Expressions Using Properties **209**

Lesson 20 Guided Practice

Practice: Evaluating Expressions Using Properties

Study the example below. Then solve problems 17–19.

Example

Jackson says that he can use more than one number property to evaluate this expression:

$$\frac{1}{4} \times (84 \times 16)$$

Look at how you could use different properties to evaluate.

Order of Operations	Associative Property	Commutative and Associative Property
$\frac{1}{4} \times (84 \times 16)$	$\frac{1}{4} \times (84 \times 16)$	$\frac{1}{4} \times (84 \times 16)$
$\frac{1}{4} \times 1{,}344$	$\left(\frac{1}{4} \times 84\right) \times 16$	$\frac{1}{4} \times (16 \times 84)$
336	21×16	$\left(\frac{1}{4} \times 16\right) \times 84$
	336	4×84
		336

Solution <u>Jackson used the order of operations, associative, and commutative and associative property to find the same solution, 336.</u>

> In the last example, Jackson first used the commutative property to change the order of the numbers and then used the associative property to change the grouping.

> **Pair/Share**
> Which method for evaluating do you like best? Why?

---

17 Selena is making fruit punch for a party. There are 50 ounces of pineapple juice and 15 ounces of cranberry juice in one batch of punch. Selena decides to triple the recipe. Write and evaluate an expression to show how many ounces of juice Selena will need.

Show your work.

> Which number properties can you use?

> **Pair/Share**
> How is your expression similar to your classmates' expressions? How is it different?

Solution _____

210　Lesson 20 Evaluate Expressions Using Properties

18 Write and evaluate a numerical expression for "18 multiplied by the difference of 9.6 and 7.9." What number property could you use to evaluate your expression?

Show your work.

What operations do you need to use to solve?

Pair/Share
How do you know which grouping symbols to use in an expression?

Solution _____

---

19 Which expression has the same value as $\frac{1}{2} \times (26 + 3)$?

Circle the letter of the correct answer.

A $\left(\frac{1}{2} \times 26\right) + 3$

B $\frac{1}{2} \times (3 + 26)$

C $26 \times \left(\frac{1}{2} + 3\right)$

D $\frac{1}{2} \times 26 + 3$

What do the parentheses tell you?

Jonas chose **C** as the correct answer. How did he get that answer?

_____
_____
_____
_____
_____

Pair/Share
Is there another number property that can be used to evaluate the expression?

©Curriculum Associates, LLC   Copying is not permitted.   Lesson 20 Evaluate Expressions Using Properties

Lesson 20 👤 **Independent Practice**

Practice > Evaluating Expressions Using Properties

Solve the problems.

1 Which equation shows equivalent expressions using the associative property?

  A $5 \times (7 - 2)$ and $(5 \times 7) - 2$

  B $12 \div (2 + 3)$ and $(12 \div 2) + 3$

  C $22 + (18 + 11)$ and $(22 + 18) + 11$

  D $27 \div (9 \div 3)$ and $(27 \div 9) \div 3$

2 Liam is evaluating expressions. For which expression can he use the commutative property to reorder all three values? Circle the letter for all that apply.

  A $5.6 \times 5.3 - 4.7$

  B $18.9 + 9.7 + 2.4$

  C $20 \div 2.4 + 6.6$

  D $5.7 \times 3.2 \times 12.4$

  E $27.8 - 19.1 + 11.3$

3 Helena earns $40 each month by helping her elderly neighbor. Of that $40, she spends $25 and saves the rest. In 6 months, how much money will Helena have saved?

  Which expression represents the problem? Circle the letter for all that apply.

  A $(40 - 25) \times 6$

  B $6 \times (40 + 25)$

  C $6 + 25 + 40$

  D $40 - 25 \times 6$

  E $6 \times (40 - 25)$

4 Which expression represents the statement "divide 12 by the difference of 4 and 2"?

A $\dfrac{12}{4+2}$

B $(12 - 4) \div 2$

C $\dfrac{12}{4} - \dfrac{12}{2}$

D $12 \div (4 - 2)$

5 Neil has 7 apps on his phone. Pilar has 4 more than 3 times the number of apps Neil has. Write a numerical expression for the number of apps Pilar has.

6 Look at the set of expressions shown below. Compare and evaluate the expressions. Which two expressions have the same value? Why?

$3 + 2 \times 5$ $\qquad\qquad (3 + 2) \times 5 \qquad\qquad 5 \times (2 + 3)$

Show your work.

Answer _____

✓ **Self Check** Go back and see what you can check off on the Self Check on page 193.

Lesson 21   Introduction
Analyze Patterns and Relationships

Use What You Know

Previously you learned to identify and continue numerical patterns. Now you will describe the relationship between two patterns. Take a look at this problem.

> Maria is working at the snack stand at a basketball game. Each frozen yogurt costs $3, and each sandwich costs $6. Create a table to show the costs for buying 0, 1, 2, 3, 4, 5, or 6 frozen yogurts. Create another table to show the costs for the same number of sandwiches. How does the cost of a given number of frozen yogurts compare to the cost of the same number of sandwiches?

a. What is the cost for buying 1 frozen yogurt? _____

b. What do you add to the cost of 1 frozen yogurt to get the cost of 2 frozen yogurts?

_____

c. Complete the table to show the cost for each number of frozen yogurts.

Number of Yogurts	0	1	2	3	4	5	6
Cost ($)							

d. What do you add to the cost of 1 sandwich to get the cost of 2 sandwiches?

_____

e. Complete the table to show the cost for each number of sandwiches.

Number of Sandwiches	0	1	2	3	4	5	6
Cost ($)							

f. How does the cost of sandwiches compare to the cost of the same number of yogurts?

_____

_____

▶▶ Find Out More

The costs of frozen yogurts and the costs of sandwiches form numerical patterns.

Cost of Frozen Yogurts ($)

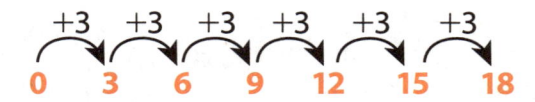

+3 +3 +3 +3 +3 +3
0   3   6   9   12 15 18

Cost of Sandwiches ($)

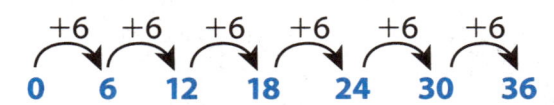

+6 +6 +6 +6 +6 +6
0   6   12 18 24 30 36

You can use a table to write ordered pairs. An **ordered pair** is a pair of numbers. Ordered pairs could help you see relationships between **corresponding terms** in two patterns. For these ordered pairs, the first number is the cost of a certain number of frozen yogurts. The second number is the cost of the same number of sandwiches.

Cost of Yogurts ($)	Cost of Sandwiches ($)	Ordered Pairs
0	0	(0, 0)
3	6	(3, 6)
6	12	(6, 12)
9	18	(9, 18)
12	24	(12, 24)
15	30	(15, 30)
18	36	(18, 36)

There is a relationship between the two numbers of these ordered pairs. Here, the second number is always twice the first number.

▶ Reflect

1 What if the ordered pairs where written as (0, 0), (6, 3), (12, 6), (18, 9), (24, 12), (30, 15), (36, 18). Would this change the relationship between the first number and the second number? Explain.

_____

_____

_____

Lesson 21 Analyze Patterns and Relationships **215**

Lesson 21 Modeled and Guided Instruction

Learn About: Comparing Two Numerical Patterns

Read the problem below. Then explore how to identify relationships between two numerical patterns.

> In Level 1 of a game, you earn 2 points for each correct answer. In Level 2, you earn 6 points for each correct answer. Compare the number of points in Level 2 to the number of points in Level 1 if you correctly answer 0, 1, 2, 3, 4, 5, or 6 questions.

▶ **Picture It** You can use a picture to help find each pattern.

Level 1
Each correct answer earns **2 points**.

+2 +2 +2 +2 +2 +2
0 2 4 6 8 10 12

Level 2
Each correct answer earns **6 points**.

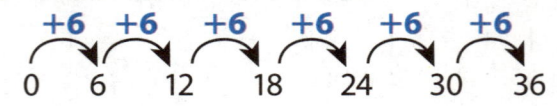

+6 +6 +6 +6 +6 +6
0 6 12 18 24 30 36

▶ **Model It** Use a table to show the number of points you get for correct answers in each level of the game.

Write ordered pairs. Have the first number be from the first pattern. Then have the second number be the corresponding number from the second pattern.

Points in Level 1	Points in Level 2	Ordered Pairs
0	0	(0, 0)
2	6	(2, 6)
4	12	
6	18	
8	24	
10	30	
12	36	

The total number of points in Level 1 increases by 2 for each correct answer. The total number of points in Level 2 increases by 6 for each correct answer.

Lesson 21 Analyze Patterns and Relationships

Connect It Now you will solve the problem from the previous page by looking at the ordered pairs.

2 Look at *Picture It* on the previous page. Describe how the total number of points changes with each correct answer in Level 1 and Level 2.

Level 1 rule: _____

Level 2 rule: _____

3 Complete the table on the previous page.

4 For each ordered pair, how does the second number compare to the first number?

_____

5 Suppose the game has a third level. You get 9 points for each correct answer in Level 3. Explain how you could figure out how the points in Level 3 compare with the corresponding points in Level 2.

_____
_____
_____
_____

Try It Use what you just learned about comparing two number patterns to solve this problem. Show your work on a separate sheet of paper.

6 School magnets cost $4, and shirts cost $24. Write a pattern for the costs of 0, 1, 2, 3, 4, and 5 magnets and a second pattern for the costs of 0, 1, 2, 3, 4, and 5 shirts. How do the corresponding terms of the two patterns compare?

_____
_____
_____

Lesson 21 Analyze Patterns and Relationships **217**

Lesson 21 Modeled and Guided Instruction

Learn About: Graphing Ordered Pairs

Read the problem below. Then explore plotting corresponding terms of number patterns on a graph.

> The scouts are making model vehicles. They have a choice of making a model plane or a model boat.
> - The materials for the plane cost $2.
> - The materials for the boat cost $4.
>
> Write and graph ordered pairs to compare the cost of making one or more boats to the cost of making the same number of planes.

▶ **Picture It** You can use a picture to find the costs of making various numbers of each model.

Each pattern of numbers below shows the cost of making 0, 1, 2, 3, and 4 models.

Planes

+2 +2 +2 +2
0   2   4   6   8

Boats

+4 +4 +4 +4
0   4   8   12   16

▶ **Model It** You can use a table to help understand the problem.

List the cost of materials for the planes and boats in a table. Then write the corresponding costs as ordered pairs.

Cost of Planes ($), x	Cost of Boats ($), y	Ordered Pairs (x, y)
0	0	(0, 0)
2	4	(2, 4)
4	8	(4, 8)
6	12	(6, 12)
8	16	(8, 16)

Lesson 21 Analyze Patterns and Relationships

Connect It Now you will solve the problem from the previous page by graphing ordered pairs.

7 Explain how to write ordered pairs using the numbers in each pattern.

_____

_____

8 Plot the ordered pairs on the graph to the right. The first number tells the location along the horizontal axis. The second number tells the location along the vertical axis.

The point (2, 4) is plotted for you.

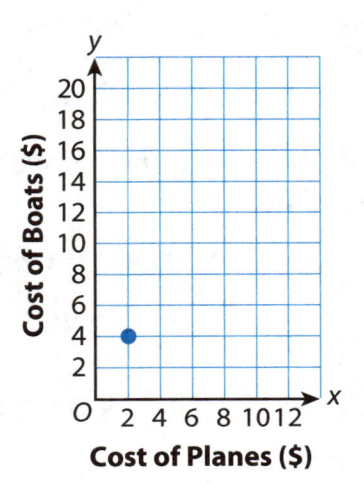

9 How do the corresponding terms of the patterns compare?

_____

_____

10 Suppose you connect the points. How would this look?

_____

11 What directions would you give to someone to get from one point to the next point to the right on the graph? How do your directions relate to the rules for the patterns?

_____

_____

_____

▶ **Try It** Use what you just learned about using graphs to compare two patterns to solve this problem. Show your work on a separate sheet of paper.

12 Consider the two patterns below. Start each pattern with 0.

Pattern A: add 1   **Pattern B:** add 3

Write five ordered pairs made up of corresponding terms from the two patterns. Plot the points on the graph to the right. Describe the relationship between the two patterns.

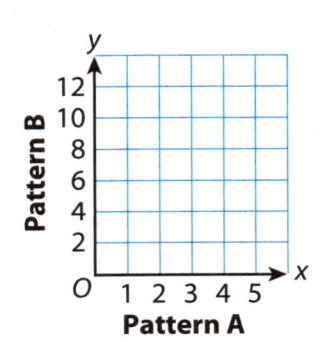

_____

_____

Lesson 21 Analyze Patterns and Relationships **219**

Lesson 21 Guided Practice

Practice: Analyzing Patterns and Relationships

Study the example below. Then solve problems 13–15.

Example

Look at the following two number patterns.

Pattern A: 6, 5, 4, 3, 2, 1, 0

Pattern B: 24, 20, 16, 12, 8, 4, 0

What is the relationship between corresponding terms in the two patterns?

Look at how you could show your work using ordered pairs.

The first number is a term from Pattern A. The second number is the corresponding term from Pattern B.

Ordered pairs: (6, 24), (5, 20), (4, 16), (3, 12), (2, 8), (1, 4), (0, 0)

Solution _Each term in Pattern B is four times the corresponding term in Pattern A._

The student wrote ordered pairs to identify a relationship between corresponding terms.

Pair/Share
How are these patterns different from other patterns in this lesson?

---

13 One pattern starts at 0 and has the rule "add 8." Another pattern starts at 0 and has the rule "add 4." Write each pattern of numbers. How do the corresponding terms in the patterns compare?

Show your work.

How do I generate the patterns?

Solution _____

Pair/Share
Does it matter how many terms you write for each pattern?

14 Identify the pattern in each column of the table. Complete the *x* and *y* columns of the table. Use those columns to write ordered pairs in the last column. Describe the relationship between corresponding terms in the patterns.

Show your work.

x	y	Ordered Pairs (x, y)
4	1	
8	2	
12	3	

What is the rule for each pattern?

Solution _____

Pair/Share
Find the difference between the numbers in each ordered pair. Do you see another pattern?

15 The ordered pairs (2, 12), (3, 18), and (4, 24) are formed by corresponding terms in two patterns. How do the values of the second numbers compare to the values of the first numbers? Circle the letter of the correct answer.

A 10 more

B 2 times as much

C $\frac{1}{6}$ times as much

D 6 times as much

Mike chose **C** as the correct answer. How did he get that answer?

_____

_____

_____

What rule works for all the ordered pairs?

Pair/Share
Does Mike's answer make sense?

Lesson 21 Analyze Patterns and Relationships **221**

Lesson 21 Independent Practice

Practice: Analyzing Patterns and Relationships

Solve the problems.

1 What rule could be used to create the pattern 9, 18, 27, 36, 45, 54, . . . ?

A   multiply each term by 2 to get the next term

B   multiply each term by 9 to get the next term

C   add 9 to get the next term

D   add 3 to get the next term

2 Look at the patterns below. Choose *True* or *False* for each statement.

Pattern A: 3, 6, 9, 12, 15, 18, . . .

Pattern B: 18, 36, 54, 72, 90, 108, . . .

a. The rule for Pattern A is "multiply by 2." ☐ True ☐ False

b. The rule for Pattern B is "add 18." ☐ True ☐ False

c. Each term in Pattern A is 6 times the corresponding term in Pattern B. ☐ True ☐ False

d. Each term in Pattern B is 3 times the corresponding term in Pattern A. ☐ True ☐ False

3 Tickets at a play cost $2 for students and $8 for adults. Jason creates two patterns to compare the costs. He writes ordered pairs in the form (student cost, adult cost) for the corresponding numbers of tickets. Which ordered pair could be on Jason's list of ordered pairs? Circle the letter for all that apply.

A   (8, 2)

B   (10, 40)

C   (4, 10)

D   (10, 16)

E   (6, 24)

4 Use the rules below to create two patterns each starting with 0. Then describe the relationship between corresponding terms of the two patterns.

Pattern A: add 3   **Pattern B:** add 12

Show your work.

_____

_____

5 Use the rules "add 2" and "add 5" to create two patterns each beginning with 0. Use the patterns to complete the table in Part A.

Part A

Add 2, x	Add 5, y	Ordered Pairs (x, y)

Part B Graph the ordered pairs and connect the points. How does this look?

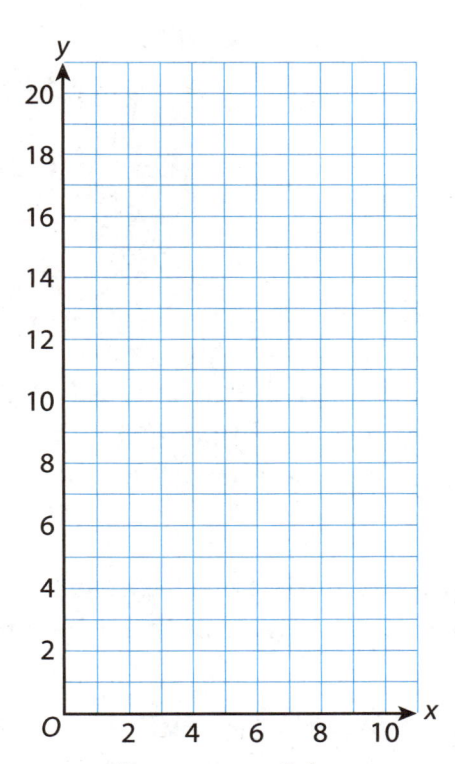

_____

_____

Part C Describe the relationship between the corresponding terms of the patterns.

_____

_____

✓ **Self Check**   Go back and see what you can check off on the Self Check on page 193.

©Curriculum Associates, LLC   Copying is not permitted.   Lesson 21 Analyze Patterns and Relationships   **223**

Unit 3 MATH IN ACTION

Introduction
Expressions, Patterns, and Relationships

SMP1 Make sense of problems and persevere in solving them.

Study an Example Problem and Solution

In this lesson, you will look at expressions, patterns, and relationships between numbers to solve real-world problems. Look at this problem and one solution.

Intermission Snacks

Brandi is working with a theater group to plan their next show. She needs to decide what snacks to sell during intermission. Brandi lists the costs to buy or make each item. The selling price will be about 2 or 3 times the cost.

Our Costs

Popcorn	$0.75
Trail Mix	$1.25
Hot Dog	$1.50
Sandwich	$2.00
Small Pizza	$2.50

Choose 2 food items. Decide on a selling price for each. Use ordered pairs to compare the amount of money they will take in by selling 0, 1, 2, 3, 4, and 5 of each item. Describe how the numbers in each ordered pair are related.

Read the sample solution on the next page. Then look at the checklist below. Find and mark parts of the solution that match the checklist.

✏️ Problem-Solving Checklist

- ☐ Tell what is known.
- ☐ Tell what the problem is asking.
- ☐ Show all your work.
- ☐ Show that the solution works.

a. **Circle** something that is known.
b. **Underline** something that you need to find.
c. **Draw a box around** what you do to solve the problem.
d. **Put a checkmark** next to the part that shows the solution works.

224 Unit 3 Math in Action Expressions, Patterns, and Relationships

Brandi's Solution

> Hi, I'm Brandi. Here's how I solved this problem.

▷ **I know** what each food item costs. I'll choose popcorn and trail mix because they seem like good snacks.

▷ **I also know** that the selling price should be about 2 or 3 times the cost. I can multiply each cost by both 2 and 3. I can use either of those products. Or I can find a number somewhere between the two products.

Item	Cost	Cost × 2	Cost × 3	My Selling Price
Popcorn	$0.75	$1.50	$2.25	$2.00
Trail Mix	$1.25	$2.50	$3.75	$3.00

> I used a table so the money amounts are clearly labeled and easy to see.

▷ **I'll check** that the selling price is between 2 and 3 times the cost.
$2.00 > $1.50   and $2.00 < $2.25
$3.00 > $2.50   and $3.00 < $3.75

▷ **I can make a** table to find the price to buy 0, 1, 2, 3, 4, and 5 of each item.

Number of items	0	1	2	3	4	5
Price of Popcorn	0	2	4	6	8	10
Price of Trail Mix	0	3	6	9	12	15
Ordered Pair	(0, 0)		(2, 3)	(4, 6)	(6, 9)	(8, 12)	(10, 15)

> I decided to use whole-dollar amounts so it will be easier to compare numbers.

▷ **I'll look at the ordered pairs** to see how the numbers are related. It looks like each second number is $1\frac{1}{2}$ times the first number.

$2 \times \frac{3}{2} = 3$ $4 \times \frac{3}{2} = 6$ $6 \times \frac{3}{2} = 9$ $8 \times \frac{3}{2} = 12$ $10 \times \frac{3}{2} = 15$

> $1\frac{1}{2}$ is the same as $\frac{3}{2}$.

I'm right. The price of trail mix in each ordered pair is $1\frac{1}{2}$ times the price of popcorn.

Unit 3 Math in Action Modeled and Guided Instruction

Try Another Approach

There are many ways to solve problems. Think about how you might solve the Intermission Snacks problem in a different way.

Intermission Snacks

Brandi is working with a theater group to plan their next show. She needs to decide what snacks to sell during intermission. Brandi lists the costs to buy or make each item. The selling price will be about 2 or 3 times the cost.

Our Costs	
Popcorn	$0.75
Trail Mix	$1.25
Hot Dog	$1.50
Sandwich	$2.00
Small Pizza	$2.50

Choose 2 food items. Decide on a selling price for each. Use ordered pairs to compare the amount of money they will take in by selling 0, 1, 2, 3, 4, and 5 of each item. Describe how the numbers in each ordered pair are related.

▶ **Plan It** Answer these questions to help you start thinking about a plan.

A. Look at the list of items. What would be a reasonable price to charge for each of them?

_____

_____

B. Will you use whole-dollar amounts as your selling prices? Why or why not?

_____

_____

226 Unit 3 Math in Action Expressions, Patterns, and Relationships

▶ **Solve It** Find a different solution for the Intermission Snacks problem. Show all your work on a separate sheet of paper.

You may want to use the problem-solving tips to get started.

Problem-Solving Tips

- **Tools** You may want to use . . .
  - a table.
  - a graph.

- **Word Bank**

ordered pair	times	pattern
corresponding	rule	relationship
terms		

- **Sentence Starters**
  - In each ordered pair _____
  - I can compare _____

Problem-Solving Checklist

Make sure that you . . .
☐ tell what you know.
☐ tell what you need to do.
☐ show all your work.
☐ show that the solution works.

▶ **Reflect**

Use Mathematical Practices As you work through the problem, discuss these questions with a partner.

- **Reason Mathematically** How can you decide what is a reasonable selling price for an item?
- **Repeated Reasoning** What is the pattern in the prices for 0, 1, 2, 3, 4, and 5 items?

Unit 3 Math in Action — Guided Practice

Discuss: Models and Strategies

Read the problem. Write a solution on a separate sheet of paper. Remember, there can be lots of ways to solve a problem!

Printing Costs

Brandi is checking prices for printing programs for the play. The program will be printed as a booklet. Here is information from a printer's website.

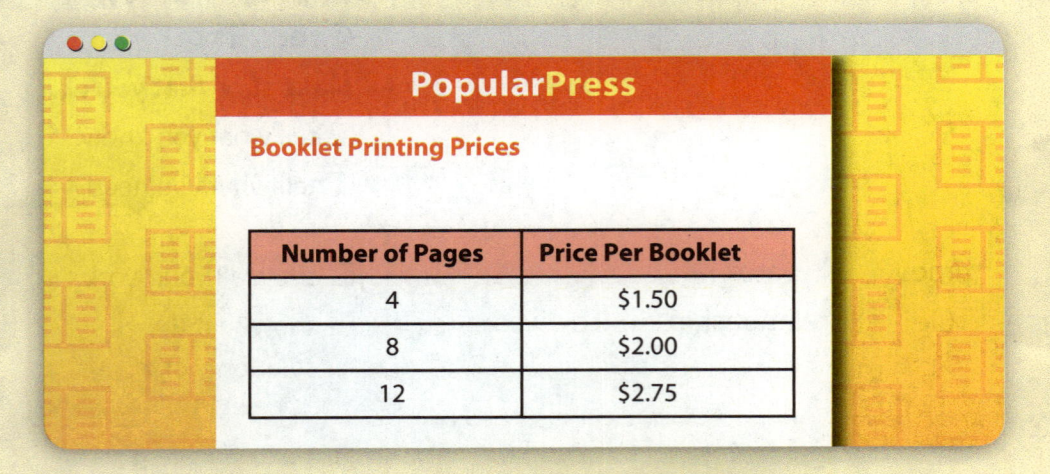

PopularPress
Booklet Printing Prices

Number of Pages	Price Per Booklet
4	$1.50
8	$2.00
12	$2.75

Brandi is not sure how many pages to have in the program. She also needs to know how many programs to print. Her decision will be partly based on price.

What will it cost to print different numbers of booklets with different numbers of pages?

▶ **Plan It and Solve It** Find a solution to the Printing Costs problem.

Compare the costs to print booklets with different numbers of pages.
- Choose two different size booklets.
- Find the cost of printing 1, 2, 3, 4, and 5 booklets with these number of pages.
- Make a graph to compare the costs.
- What is the relationship between the two numbers in each ordered pair? Write a statement to describe this relationship.

Problem-Solving Tips

- **Questions**
  - What will you use as a scale on your graph?
  - Look at the numbers in each ordered pair. What operation can you use to describe the relationship between these numbers?

- **Word Bank**

ordered pair	patterns	table
corresponding	graph	coordinates
terms		

Problem-Solving Checklist
Make sure that you . . .
☐ tell what you know.
☐ tell what you need to do.
☐ show all your work.
☐ show that the solution works.

▶ **Reflect**

Use Mathematical Practices As you work through the problem, discuss these questions with a partner.

- **Reason Mathematically** When you describe the relationship between corresponding terms, how do you know what operation to use?
- **Make Sense of Problems** How can the information in the problem help you choose the number of pages to use?

Unit 3 Math in Action   **Independent Practice**

Persevere ▶ On Your Own

Read the problems. Write a solution on a separate sheet of paper. Remember, there are many different ways to solve a problem.

Snack Discounts

Brandi does not want to have much leftover food when the play is over. She will check what is left about halfway through the intermission. If there is a lot of food left, Brandi will offer a discount.

Here are two discounts Brandi is considering.

- Pay $\frac{1}{2}$ of the total price if you buy two items.
- Pay $\frac{1}{4}$ of the total price if you buy four items.

Here is Brandi's regular price list.

Price List	
Popcorn	$2
Trail Mix	$3
Hot Dog	$4
Sandwich	$6
Small Pizza	$8

How much will it cost to buy different food items with the discounts?

▶ **Solve It** Help Brandi get an idea of what customers will pay for food with the discounts.

- Choose two different combinations of items for each discount.
- Write an expression that can be used to find the total costs.
- Evaluate the expressions and show the total cost for each combination.

▶ **Reflect**

Use Mathematical Practices After you complete the task, choose one of these questions to discuss with a partner.

- **Reason with Numbers** What do the numbers in your expressions mean in the context of the problem?

- **Repeated Reasoning** How did you calculate the fraction of a price? Explain.

Play Programs

Now Brandi is considering the size of the program.

Printing Costs
- $1.50 for each 4-page booklet
- $2.00 for each 8-page booklet

Brandi will double the printing cost to find the *selling price* of a program. The money she makes selling programs will be used to pay the printing costs. The money she has left is her *profit*.

Brandi needs to decide whether to make 4-page or 8-page programs. She thinks that she can sell more programs at a lower price. But she thinks she might make more profit with higher-price programs. Here are Brandi's sales estimates.

- I can sell 280 to 300 of the 4-page programs.
- I can sell 250 to 280 of the 8-page programs.

About how much profit might Brandi make selling programs? How many pages should the programs be?

▶ **Solve It** Help Brandi decide whether to make 4-page or 8-page programs.

- Choose a number of 4-page programs and a number of 8-page programs from Brandi's sales estimates.
- Use these numbers to write and evaluate expressions that show how much profit Brandi will make after paying for the programs to be printed.
- Tell which page length Brandi should use and support your recommendation.

▶ **Reflect**

Use Mathematical Practices After you complete the task, choose one of these questions to discuss with a partner.

- **Make an Argument** How did you support your recommendation?
- **Reason Mathematically** How did you choose the number of 4-page and the number of 8-page programs?

Unit 3 Assessment
Interim Assessment

Solve the problems.

1. Bodi and 4 friends each bought a $15 ticket to the Pro Football Hall of Fame. Bodi had a coupon for $10 off the total cost of the tickets. Which expression can be used to find the total amount paid by Bodi and his friends?

   A (5 × 15) − 10

   B 5 × (15 − 10)

   C (15 − 10) × 5

   D (15 + 10) × 5

2. Zhen and Justin made two different number patterns that each started at 0. Zhen wrote terms using the rule "add 18." Justin wrote terms using the rule "add 6." What is the relationship between the two patterns?

   A The terms in Zhen's pattern are 3 greater than the corresponding terms in Justin's pattern.

   B The terms in Zhen's pattern are 3 less than the corresponding terms in Justin's pattern.

   C The terms in Zhen's pattern are 3 times the corresponding terms in Justin's pattern.

   D The terms in Zhen's pattern are $\frac{1}{3}$ times the corresponding terms in Justin's pattern.

3. Emery practices the piano a total of 150 minutes Monday through Friday. He practices a total of 25 minutes on the weekends. In 5 weeks, how many minutes does Emery practice the piano?

   Which expression represents the problem?

   A 150 + 25 × 5

   B 5 × 150 + 25

   C (150 × 5) + 25

   D 5 × (150 + 25)

4. Which expression has the value of 56?

   A 8 + 4 × 6

   B 4 × 6 + 8

   C (8 × 4) + 6

   D 4 × (6 + 8)

5 The table shows the first four terms for Pattern A. The rule for Pattern A is "add 2." Pattern B, begins at 0. Pattern B uses a rule that adds the same number to each term. Each term in Pattern B is 10 times the corresponding term in Pattern A.

Pattern A	Pattern B
0	0
2	____
4	____
6	____

What is the rule for Pattern B starting at 0?

A add 10

B add 20

C multiply by 10

D multiply by 20

6 Which expression uses the distributive property to show an expression that has the same value as 7 × (5 + 6)?

A (7 × 5) + (7 × 6)

B (6 × 7) + (6 × 5)

C (5 + 7) × (5 + 6)

D (7 + 6) × (7 + 5)

7 Which expression represents the statement "subtract 13 from the product of 4 and 7"?

A (13 − 4) × (13 − 7)

B (13 × 4) − (13 × 7)

C 4 × 7 − 13

D 13 − 4 × 7

Unit 3 Interim Assessment continued

Performance Task

Answer the questions and show all your work on separate paper.

Your friend Sophie has asked you to check her math homework. She sends you a text message with the expressions she evaluated and the values she found. Unfortunately, she didn't include any parentheses in her message.

Copy each equation below. If the equation is true as written, write a check mark (✓) next to it. If the equation is not true, insert parentheses to make the equation true.

$6 + 5 \times 3 = 33$ $12 \div 2 + 4 = 10$ $15 - 3 \times 2 = 9$

$6 + 5 \times 3 = 21$ $12 \div 2 + 4 = 2$ $15 - 3 \times 2 = 24$

Then write a message to Sophie explaining why she needed to include the parentheses in order for you to check her homework.

Checklist
Did you . . .
- ☐ check your calculations?
- ☐ use order of operations?
- ☐ use complete sentences?

▶ **Reflect**

Use Mathematical Practices After you complete the task, choose one of the following questions to answer.

- **Be Precise** How did you decide where to insert the parentheses in each equation?

- **Use Structure** Consider the expression $6 + 7 \times 5$. Would you rewrite it as $6 + (7 \times 5)$? Why or why not?

Unit 4
Measurement and Data

Let's learn about finding volume using different units of measurement.

Real-World Connection You can talk about the same measurement in different ways. You might say that 52 buckets of sand will fill your backyard sandbox. The clerk at the hardware store where you bought the sand might say the sandbox needs 24 cubic feet of sand. If it's summertime, you could fill your sandbox with water instead of sand. Then you might say that it will hold about 180 gallons of water. All of these different measurements are ways to talk about the volume of the sandbox.

In This Unit You will find the volume of different kinds of figures using different units of measurement. You will use what you know about multiplication to find the volume of figures using a formula. You will also solve word problems involving different units of measurement and create line graphs of collected data and then interpret the data.

✓ Self Check

Before starting this unit, check off the skills you know below. As you complete each lesson, see how many more skills you can check off!

I can:	Before this unit	After this unit
convert from one measurement unit to another, for example: 4 feet = 48 inches.	☐	☐
make and interpret a line graph that represents collected data.	☐	☐
find the volume of a solid figure by counting unit cubes.	☐	☐
find volume by using a formula.	☐	☐
find the volume of composite figures.	☐	☐

235

Lesson 22 Introduction
Convert Measurement Units

Use What You Know

You have worked with measurement units in earlier grades. Now you will convert between different units in the same measurement system.

Is the number of cups in 5 gallons greater than or less than the number of gallons?

1 gallon

1 cup

a. Circle the greater amount in each row:

   1 gallon   or   1 cup

   2 gallons   or   2 cups

   5 gallons   or   5 cups

b. If you pour 5 gallons of water into 1-cup containers, would you need more than or fewer than 5 containers?

   _____

c. Is the number of cups in 5 gallons greater than or less than the number of gallons?

   _____

▶▶ Find Out More

You measure for many different reasons. You might measure to find how long or tall something is, how much liquid something holds, or how much something weighs.

You can choose different units when you measure. Think about your height. You could measure your height in inches or feet. Your height does not change if you are measured in inches instead of feet. It is just recorded using different units.

Look at the picture at the right. **1 gallon = 16 cups**

The same amount of liquid could also be measured in quarts. Quarts are smaller than gallons.

 1 gallon = 4 quarts

Quarts are larger than cups. **1 quart = 4 cups**

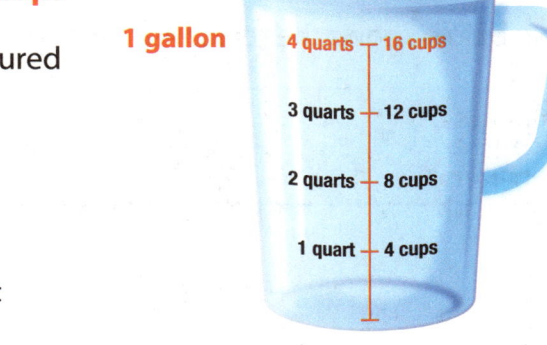

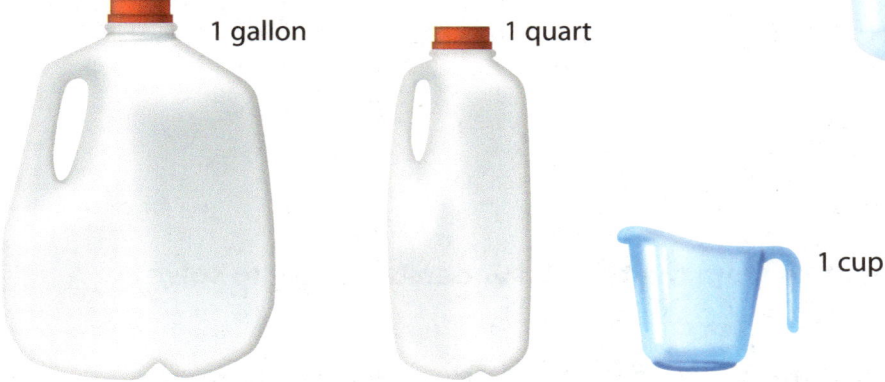

1 gallon   4 quarts — 16 cups
            3 quarts — 12 cups
            2 quarts — 8 cups
            1 quart — 4 cups

1 gallon   1 quart   1 cup

Imagine filling the 1-gallon container using cups or quarts. You would need to fill the quart container 4 times to have enough liquid to fill the gallon container. You would need to fill the cup container 16 times to have enough liquid to fill the gallon container.

▶ Reflect

1 Describe a real-world object that can be measured using two different units. Which unit would you need more of to measure the object?

___

___

___

Lesson 22 Convert Measurement Units **237**

Lesson 22 Modeled and Guided Instruction

Learn About Converting Measurements Using Multiplication

Read the problem below. Then explore different ways to convert measurement units.

> How many meters are in 3.5 kilometers?

▶ **Model It** You can use a table to help understand the problem.

The table below shows the relationship between meters and kilometers.

kilometers	1	2	3	4	5	6
meters	1,000	2,000	3,000	4,000	5,000	6,000

▶ **Solve It** Use the information from the table to understand how to solve the problem.

The pattern in the table shows that the number of meters is always 1,000 times the number of kilometers.

kilometers	1	2	3	3.5	4	5	6
meters	1,000	2,000	3,000		4,000	5,000	6,000

To find the number of meters in 3.5 kilometers, multiply 3.5 by 1,000.

238 Lesson 22 Convert Measurement Units

▶ **Connect It** Now you will solve the problem from the previous page using unit conversions.

> 1 kilometer = 1,000 meters

2 Which is the smaller unit, meters or kilometers? _____

How do you know? _____

_____

3 What operation do you use to convert from a larger measurement unit to a smaller measurement unit? _____

4 3.5 kilometers = _____ meters

Write your answer in the table on the previous page.

5 Use what you learned about the relationship between meters and kilometers to complete the table below.

kilometers	0.8	1	1.85	2	2.03	3
meters		1,000		2,000		3,000

6 How many meters are in k kilometers? _____

7 There are 3 feet in 1 yard. Explain how you decide whether to multiply or divide by 3 if you need to convert yards to feet. _____

_____

_____

▶ **Try It** Use what you just learned about converting measurement units to solve these problems. Show your work on a separate sheet of paper.

8 How many ounces are in $10\frac{1}{2}$ pounds?

> 1 pound = 16 ounces

_____

9 How many millimeters are in 9.25 centimeters?

> 1 centimeter = 10 millimeters

_____

Lesson 22 Convert Measurement Units **239**

Lesson 22 Modeled and Guided Instruction

Learn About: Converting Measurements Using Division

Read the problem below. Then explore ways to understand how to convert measurement units.

> How many quarts are equivalent to 6 cups?

▶ **Model It** You can use a table to help understand the problem.

The table below shows the relationship between cups and quarts.

quarts	1	2	3	4	5	6
cups	4	8	12	16	20	24

▶ **Solve It** Use the information from the table to understand how to solve the problem.

The pattern in the table shows that there are 4 cups in every quart.

quarts	1		2	3	4	5	6
cups	4	6	8	12	16	20	24

To find the number of quarts equivalent to 6 cups, divide 6 by 4.

Connect It Now you will solve the problem from the previous page using unit conversions.

> 1 quart = 4 cups

10 Which is a smaller unit, quarts or cups? _____

How do you know? _____

_____

11 What operation do you use to convert from a smaller measurement unit to a larger measurement unit? _____

12 6 cups = _____ quarts

Write your answer in the table on the previous page. Explain your reasoning.

_____

_____

13 Use what you learned about the relationship between cups and quarts to complete the table below.

quarts		1		2		3		4
cups	2	4	5	8	9	12	15	16

14 One gallon is equivalent to 8 pints. Describe how to convert from pints to gallons. Explain your reasoning. _____

_____

_____

Try It Use what you just learned about converting measurement units to solve these problems. Show your work on a separate sheet of paper.

15 How many liters are in 100 milliliters?

> 1 liter = 1,000 milliliters

_____

16 How many minutes are equivalent to 270 seconds?

> 1 minute = 60 seconds

Lesson 22 Convert Measurement Units

Lesson 22 Guided Practice

Practice: Converting Measurement Units

Study the example below. Then solve problems 17–19.

Example

Write 75 inches in feet and inches.

| 1 foot = 12 inches |

Look at how you could explain your work using conversions.

Feet are larger than inches, so the number of feet will be less than the number of inches. Divide.

Since there are 12 inches in 1 foot: $75 \div 12 = \frac{75}{12} = 6\frac{3}{12}$.

The whole number, 6, is the number of feet and the fraction, $\frac{3}{12}$, is $\frac{3}{12}$ foot. $\frac{1}{12}$ foot is 1 inch, so $\frac{3}{12}$ foot is 3 inches.

75 inches = 6 feet, 3 inches

Solution _6 feet, 3 inches_

> The student used division because inches are a smaller unit than feet.

> **Pair/Share**
> How can you check your answer?

---

17 How many minutes are there in 10.5 hours?

| 1 hour = 60 minutes |

Show your work.

> Which measurement unit is larger, minutes or hours?

> **Pair/Share**
> Draw a picture or make a table to support your answer.

Solution _____

242   Lesson 22 Convert Measurement Units

18 How many kilograms are equivalent to 450 grams?

1 kilogram = 1,000 grams

Show your work.

Will the number of kilograms be greater or less than 450?

Pair/Share
Did you and your partner solve the problem the same way?

Solution _____

---

19 How many millimeters are in 118 meters? Circle the letter of the correct answer.

1 meter = 1,000 millimeters

A 0.0118 millimeter

B 0.118 millimeter

C 118,000 millimeters

D 1,180,000 millimeters

Emily chose **B** as the correct answer. How did she get that answer?

_____

_____

_____

Which is a larger unit, millimeters or meters?

Pair/Share
Does Emily's answer make sense?

Lesson 22   Independent Practice

Practice > Converting Measurement Units

Solve the problems.

1 How many grams are equivalent to 75 kilograms?

> 1 kilogram = 1,000 grams

   A   0.0075 gram

   B   0.075 gram

   C   75,000 grams

   D   750,000 grams

2 How many yards and feet are equivalent to 10,000 feet?

> 1 yard = 3 feet

   A   3,333 yards, 0 feet

   B   3,333 yards, 1 foot

   C   277 yards, 28 feet

   D   30,000 yards, 0 feet

3 Write each measurement below in the table under an equivalent measure. Some of the measurements may not have an equivalent measure.

> 1 gallon = 4 quarts
> 1 quart = 2 pints
> 1 pint = 2 cups

$\frac{1}{2}$ quart   4 pints   16 cups $\frac{1}{4}$ gallon   8 pints

1 gallon	1 quart	1 pint

Use the conversion tables at the bottom of the page to solve each problem.

4 Five measurements are shown below. Write one measurement on each line to create two true equations.

300 millimeters   30 meters   3,000 meters   3 kilometers   3,000 centimeters

_____ = _____

_____ = _____

5 How many pints are equivalent to 3 quarts?

Show your work.

Answer _____ pints

6 Complete each conversion below.

Show your work.

a. 3 feet, 7 inches = _____ inches

b. 3 days = _____ hours

c. 5 pounds = _____ ounces

d. 4.6 meters = _____ centimeters

e. 7,000 meters = _____ kilometers

f. 0.85 liter = _____ milliliters

Units of Length
1 foot = 12 inches
1 yard = 3 feet
1 mile = 5,280 feet
1 meter = 100 centimeters
1 meter = 1,000 millimeters
1 kilometer = 1,000 meters

Units of Capacity
1 quart = 2 pints
1 quart = 4 cups
1 gallon = 4 quarts
1 liter = 1,000 milliliters

Units of Mass
1 kilogram = 1,000 grams

Units of Weight
1 pound = 16 ounces

Units of Time
1 day = 24 hours
1 hour = 60 minutes
1 minute = 60 seconds

✓ Self Check Go back and see what you can check off on the Self Check on page 235.

©Curriculum Associates, LLC   Copying is not permitted.   Lesson 22 Convert Measurement Units

Lesson 23   Introduction
Solve Word Problems Involving Conversions

Use What You Know

You know how to convert between different units of measure. In this lesson, you will convert measurement units to solve real-world problems.

> Ray finds a log that is 90 inches long. How long is the log in yards and inches?
>
> 1 yard = 3 feet = 36 inches
>
> ← 90 inches →

a. The length of the log is given in what unit? _____

b. The problem asks to find the length of the log in what units? _____

c. Circle the larger unit below.

    yard   inch

d. To convert yards to inches, what operation do you use? _____

    To convert inches to yards, what operation do you use? _____

e. 1 yard is equivalent to 3 feet or _____ inches.

f. Explain how to express 90 inches in yards and inches.

_____

_____

_____

_____

▶▶ Find Out More

When you convert from one unit of measure to another, you need to know the relationship between the two units.

90 inches

2 yards 18 inches

Think about the log. Instead of giving the length in yards, you could give the length in feet.

- To change yards to feet, you need to know that there are 3 feet in 1 yard. Because feet are smaller than yards, you will have more feet than yards.

- To change feet to inches, you need to know that there are 12 inches in 1 foot. Because feet are larger than inches, you will have fewer feet than inches.

A yardstick is shown below. Look at the relative sizes of 1 yard, 1 foot, and 1 inch.

Yardstick on this page is not life-sized.

▶ Reflect

1 Describe how you know whether to multiply or divide when converting one unit of measure to another.

_____

_____

_____

_____

_____

Lesson 23 Solve Word Problems Involving Conversions

Lesson 23 Modeled and Guided Instruction

Learn About > Converting Units Using Equations

Read the problem below. Then explore different ways to understand how to convert the units to solve the problem.

> Corrina is making punch for a family reunion. Her recipe calls for $2\frac{1}{2}$ cups of lemonade per batch. She plans to use 50 cups of lemonade to make 20 batches of punch for the reunion. How many gallons of lemonade does she need?
>
> 1 gallon = 16 cups

▶ **Picture It** You can use a picture to understand the relationship between cups and gallons.

1 gallon

1 cup

▶ **Model It** You can use a table to help convert from a smaller unit to a larger unit of liquid volume.

The table below shows that there are 16 cups in 1 gallon. It also shows how many cups are in 2, 3, 4, and 5 gallons.

Gallons	1	2	3	4
Cups	16	32	48	64

248 Lesson 23 Solve Word Problems Involving Conversions

Connect It Now you will solve the problem from the previous page by converting units.

2 Why do you need to convert units of measure to solve the problem?

_____

_____

_____

3 What operation do you use to convert 50 cups to gallons? Explain.

_____

_____

4 Find how many full gallons of lemonade Corrina needs. Will there be any cups left over?

_____

5 If Corrina could only buy whole gallons of lemonade at the store, how many gallons would she need to buy? _____

Explain your reasoning. _____

_____

_____

6 What if lemonade is only sold in quarts? How could you find the number of quarts Corrina would need?

| 1 quart = 4 cups |

_____

_____

Try It Use what you just learned about converting units of measure to solve this problem. Show your work on a separate sheet of paper.

7 The pet store salesman told Evan to feed his dog 112 ounces of food over 2 weeks. At that store, dog food is sold by the pound. How many pounds of food does Evan need to buy to feed his dog for 2 weeks?

| 16 ounces = 1 pound |

_____

Lesson 23 Solve Word Problems Involving Conversions   **249**

Lesson 23 Modeled and Guided Instruction

Learn About: Solving Word Problems Involving Conversions

Read the problem below. Then explore different ways to understand how to convert the units to solve the problem.

> Heather and Diego measured worms in their class compost bin. Heather measured a 3.5-centimeter worm. Diego measured a 28-millimeter worm. Who measured the longer worm?
>
> 10 millimeters = 1 centimeter

▶ **Picture It** You can use a picture to help understand the relationship between centimeters and millimeters.

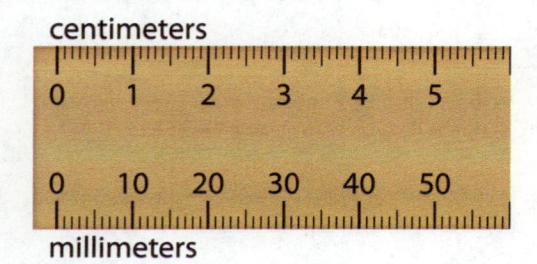

▶ **Model It** You can write equations to convert centimeters to millimeters or millimeters to centimeters.

To compare the lengths, both measurements need to be in the same unit.

Convert centimeters to millimeters:

There are 10 millimeters in each centimeter.

$3.5 \times 10 = 35$

3.5 centimeters is 35 millimeters.

Convert millimeters to centimeters:

There are 10 millimeters in each centimeter.

$28 \div 10 = 2.8$

28 millimeters is 2.8 centimeters.

Connect It Now you will solve the problem from the previous page by converting units.

8 Why do you need to convert one of the units of measure to solve the problem?

_____

_____

_____

9 Why do you multiply to convert centimeters to millimeters? _____

_____

_____

Why do you divide to convert millimeters to centimeters? _____

_____

_____

10 Who measured the longer worm? How do you know? _____

_____

_____

11 If a third student measured a worm in centimeters, how could you figure out which of the three students measured the longest worm? _____

_____

_____

_____

Try It Use what you just learned about converting units of measure to solve this problem. Show your work on a separate sheet of paper.

12 Avery's marathon time was 4 hours, 37 minutes. Evan's time was 252 minutes. Which runner had the shorter run time?

1 hour = 60 minutes

_____

Lesson 23 Solve Word Problems Involving Conversions **251**

Lesson 23 Guided Practice

Practice: Solving Word Problems Involving Conversions

Study the example below. Then solve problems 13–15.

Example

Pierre is 53 inches tall. What is his height in feet and inches?

| 1 foot = 12 inches |

Look at how you could explain your work.

Feet are longer than inches, so there are fewer feet than inches. You need to divide.

There are 12 inches in 1 foot, so divide 53 by 12.

53 ÷ 12 is 4 with a remainder of 5.

53 inches = 4 feet, 5 inches

Solution 4 feet, 5 inches

The student used the relationship between feet and inches to solve the problem.

Pair/Share
What is 53 inches in yards, feet, and inches?

13. Venell put together a model train with 25 train cars. The total length of all of the train cars is 2,000 millimeters. How many meters long is Venell's model train if there are no gaps between the cars?

| 1 meter = 1,000 millimeters |

Show your work.

Which is the larger unit, meters or millimeters?

Pair/Share
There are 100 centimeters in 1 meter. How long is the train in centimeters?

Solution _____

252 Lesson 23 Solve Word Problems Involving Conversions

14 Orla bought a 1-quart container of buttermilk to make pancakes. Her recipe uses $\frac{1}{2}$ cup of buttermilk for each pancake. How many pancakes can Orla make?

| 1 quart = 4 cups |

Show your work.

How many $\frac{1}{2}$ cups are in 1 quart?

Pair/Share
Make a table to show how much buttermilk is needed for different numbers of pancakes.

Solution _____

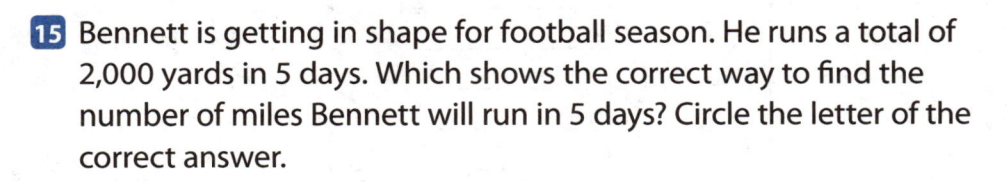

---

15 Bennett is getting in shape for football season. He runs a total of 2,000 yards in 5 days. Which shows the correct way to find the number of miles Bennett will run in 5 days? Circle the letter of the correct answer.

| 1 mile = 1,760 yards |

A 2,000 × 1,760

B 2,000 × 5 ÷ 1,760

C 2,000 ÷ 5 ÷ 1,760

D 2,000 ÷ 1,760

Jory chose **A** as the correct answer. How did he get that answer?

_____

_____

_____

_____

Does Bennett run a greater number of miles or a greater number of yards?

Pair/Share
Does Jory's answer make sense?

Lesson 23 Solve Word Problems Involving Conversions 253

Lesson 23 👤 **Independent Practice**

Practice ➤ **Solving Word Problems Involving Conversions**

Solve the problems.

1 Garrett ran 50 yards on a football field. Which shows a correct expression to find the number of feet Garrett ran?

1 yard = 3 feet

A 3 ÷ 50

B 50 ÷ 3

C 50 × 3

D 5 × 3

2 Brady practices the oboe for 315 minutes during the week. How many hours does he practice during the week?

1 hour = 60 minutes

A 5 hours

B $5\frac{1}{4}$ hours

C $5\frac{1}{2}$ hours

D $5\frac{3}{4}$ hours

3 Susan is stacking boxes on a shelf. Each box is shaped like a rectangular prism and has a length of 2 feet, a width of 15 inches, and a height of 3 inches, as shown below.

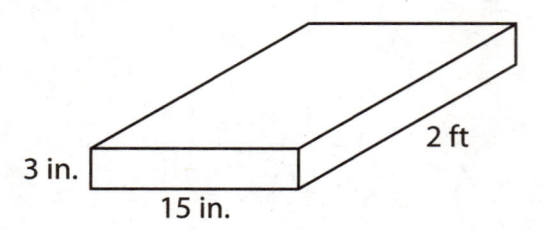

Susan will stack the boxes on top of each other, as shown in the diagram below. The space above the shelf is 1.5 feet high.

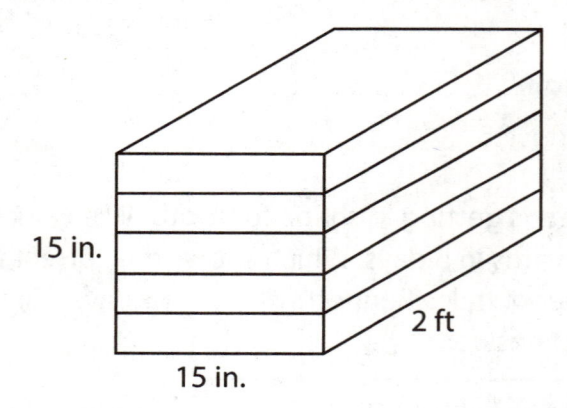

Does Susan have enough room to stack 5 boxes? Explain.

1 foot = 12 inches

_____

_____

_____

_____

_____

4 The Russell family bought 5 gallons of juice to give out at a picnic. They will put the juice into 1-cup containers. How many containers will they need if they give out all 5 gallons?

| 1 gallon = 16 cups |

Show your work.

Answer _____ containers

5 Lana's family entered a 5-kilometer race.

| 1 kilometer = 1,000 meters |

Part A Lana's dad said his average step length is about 1 meter. About how many steps will he need to take to finish the race?

Show your work.

Answer About _____ steps

Part B Lana's average step length is about 0.5 meter. How many steps will she need to take to finish the race?

Show your work.

Answer About _____ steps

✓ **Self Check** Go back and see what you can check off on the Self Check on page 235.

©Curriculum Associates, LLC   Copying is not permitted.   Lesson 23 Solve Word Problems Involving Conversions   **255**

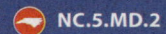

Lesson 24 — Introduction
Make Line Graphs and Interpret Data

NC.5.MD.2

Use What You Know

You have collected data before. Now you will use numerical data to create graphs. Take a look at this problem.

A movie theater owner wants to know how many tickets are sold every day for a week. She has the following results. How can she put this information into a graph to get a visual representation of the data?

Day	Sunday	Monday	Tuesday	Wednesday	Thursday	Friday	Saturday
Tickets sold	360	290	110	70	70	370	540

a. The greatest number of tickets sold was _____.

b. The least number of tickets sold was _____.

c. Below, make a bar graph showing the number of tickets sold each day.

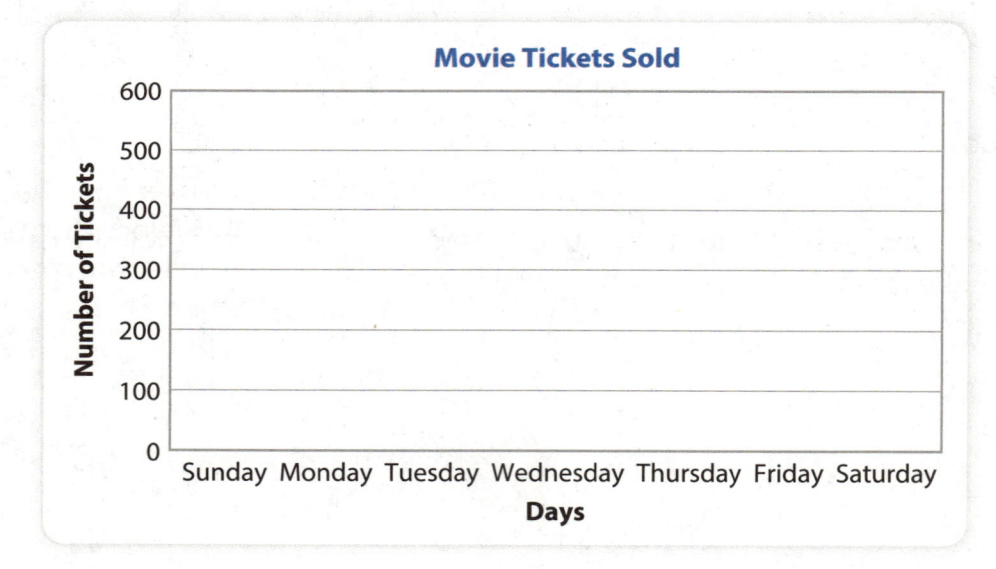

Movie Tickets Sold

(Blank bar graph: y-axis labeled "Number of Tickets" from 0 to 600 in increments of 100; x-axis labeled "Days" with Sunday, Monday, Tuesday, Wednesday, Thursday, Friday, Saturday)

d. Would it make sense to make a line plot to display this data? Explain.

_____

_____

256 Lesson 24 Make Line Graphs and Interpret Data

▶▶ Find Out More

The ticket sales data shows a change over time. You can look at the data and see whether sales went up or down from each previous day.

Not all data shows a change over time. Suppose the owner looked at how many tickets were sold for each movie on a particular day. She would get numerical results, the number of tickets sold, but they wouldn't show a change over time. Or, the owner could also ask 100 people what type of movie they like best. Such data is categorical, meaning it shows a set of data (movies) divided into groups (types of movies).

You can show data that changes over time in a **line graph**. This is a data display that shows data points connected by line segments.

Look at the bar graph you made on the previous page. It shows the days of the week in order. The bars are higher or lower based on the number of tickets sold each day. You can easily find which day had the greatest ticket sales by looking for the tallest bar.

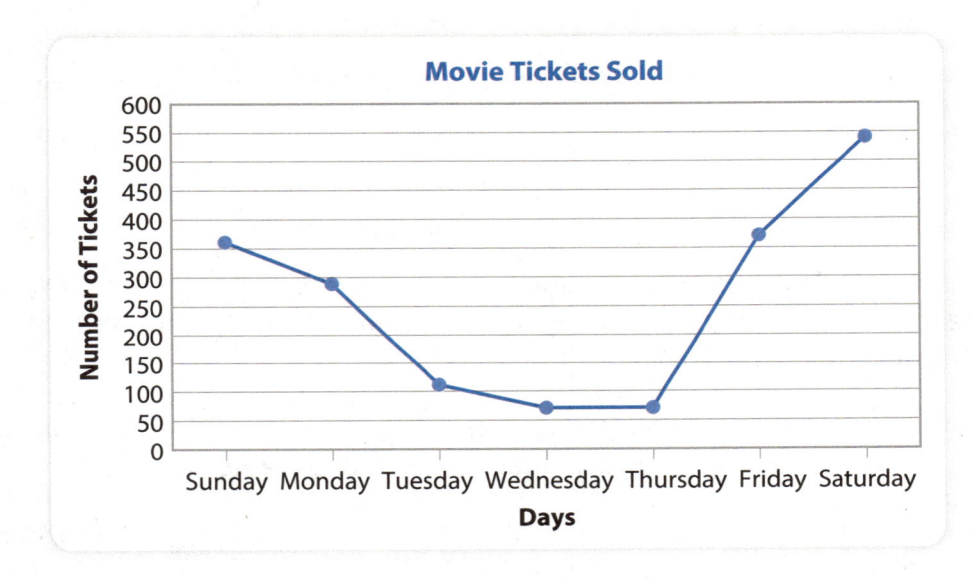

Movie Tickets Sold

The line graph uses a similar structure to show the change of the data over time. The labels on the axes show the days of the week and the number of tickets. By finding where the day of the week and the number of tickets intersect, you can plot the data. The connecting lines from each day help you see the relationship between the data for each day.

▶ Reflect

1 Would the same type of line graph make sense if the theater wanted to track sales month to month? Explain.

___

___

___

Lesson 24 Make Line Graphs and Interpret Data **257**

Lesson 24 Modeled and Guided Instruction

Learn About Types of Survey Questions

Read the problem below. Then explore how to create a survey question that will produce a specific type of data.

> Will manages a restaurant. He wants to ask his staff some questions about the restaurant. Which question will result in data that shows a change over time?
> - Question 1: What is the most popular dish you served?
> - Question 2: How many tables were seated each hour?
> - Question 3: How many children were seated in your section?

▶ **Explain It** You can examine types of data sets.

Categorical data represents characteristics or states of being. Some examples include: types of animals (dog, cat, mouse); age group (adult, child, teen); and states (Alaska, California, North Carolina).

Numerical data can be counted or measured to give a numerical result. Some examples include height, temperature, attendance, test scores, or items sold.

Change over time data is numerical data that shows a change over a specified period of time. The time period could be as short as a few seconds or as long as several centuries. The data should represent regular intervals, but the length of the time period is up to the person collecting the data as well as what data is available.

▶ **Model It** You can create possible answers to each question to understand the problem.

Question 1 What is the most popular dish you served?

Possible answers burger, spinach salad, ravioli, chicken salad, chowder, burger

Question 2 How many tables were seated each hour?

Possible answers 6:00 P.M.: 5 tables; 7:00 P.M.: 8 tables; 8:00 P.M.: 5 tables; 9:00 P.M.: 3 tables; 10:00 P.M.: 1 table

Question 3 How many children were seated in your section?

Possible answers 0, 3, 1, 5, 1, 0

Connect It Now you will solve the problem from the previous page using what you have learned about types of data.

2 Which of the possible answers in *Model It* is a type of data set that is not numerical?

_____
_____

3 Which possible answers give numerical data? _____

_____
_____

4 In *Model It*, which words in Questions 2 and 3 signal that the data is going to be numerical? _____

5 Which of the possible answers shows a change over time?

_____
_____

6 Which question should the manager ask to find data that shows a change over time?

_____

Try It Use what you just learned about creating survey questions to solve these problems.

7 How could you change Question 1 in *Model It* so that the answer is data about a specific dish and shows a change over time? _____

_____
_____

8 What is another question the manager could ask to get data that is categorical?

_____
_____
_____

Lesson 24 Make Line Graphs and Interpret Data

Lesson 24 **Modeled and Guided Instruction**

Learn About: Collecting and Displaying Data That Changes Over Time

Read the problem below. Then explore how to display data that changes over time in a line graph.

> Gabrielle wants to find out whether attendance at the games of a local basketball team has increased or decreased. She tracks the total number of tickets sold each month for a whole season. The ticket sales for each month are shown below.
>
> November: 8,642; December: 7,424; January: 8,184; February: 6,154; March: 9,524
>
> How can she display this data in a line graph?

▶ Model It You can use a table to organize the values.

A table can help you organize the data. You can see the range of values, identify the beginning and the end of the time period, and find the greatest number of tickets sold.

Month	November	December	January	February	March
Attendance	8,642	7,424	8,184	6,154	9,524

▶ Model It Use the table to list what you know and to plan how to make the line graph.

- The vertical axis starts at 0 and ends at a point greater than the largest data point in the set.
- The data values are in the thousands.
- The horizontal axis is labeled with the months.

Lesson 24 Make Line Graphs and Interpret Data

Connect It Now you will use the information on the previous page to make a line graph to display Gabrielle's data.

9 What is the greatest number of tickets sold? _____

10 What scale can you use to label the lines on the vertical axis?
_____

11 Is it important to list the months in order? Explain why or why not.
_____
_____

12 Now use the data from the problem to plot the points and make a line graph.

Try It Use what you just learned about collecting and displaying data that changes over time to solve these problems.

13 Write a question for data that changes over time that you can collect data on.
_____

14 Collect the data and make a line graph to answer your question.
_____

Lesson 24 Make Line Graphs and Interpret Data

Lesson 24 Modeled and Guided Instruction

Learn About Interpreting Line Graphs

Read the problem below. Then explore using line graphs to solve a problem.

Kei makes a line graph to track the money in his savings account over one year. Does Kei end the year with more or less money than he had at the beginning?

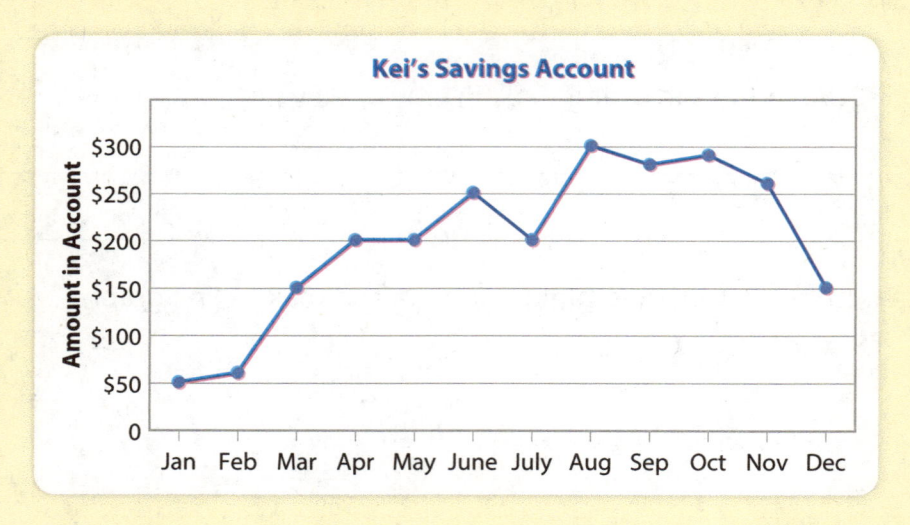

Kei's Savings Account

▶ **Model It** You can identify data points and their values.

The problem asks about the beginning and end of the year. Identify the data points for the beginning and end of the year.

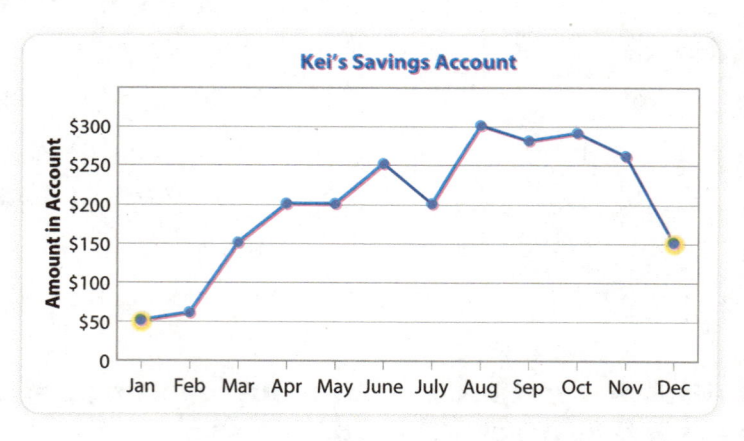

Kei's Savings Account

▶ **Picture It** You can use a picture to help understand the data in the problem.

Money in account in January: $50

Money in account in December: $150

262 Lesson 24 Make Line Graphs and Interpret Data ©Curriculum Associates, LLC Copying is not permitted.

Connect It Now you will use what you've learned on the previous page to answer the question about Kei's savings account.

15 Is the data point for December higher or lower on the graph than the point for January? _____

16 Does a higher data point mean that Kei had more or less in his savings account? Explain.

_____

_____

17 What was the amount in savings in January? _____

18 What was the amount in savings in December? _____

19 Did Kei end with more or less money than at the beginning of the year? How much more or less? Explain. _____

_____

20 How much did the amount in Kei's account change between March and June?

_____

Try It Use what you just learned about line graphs to solve these problems.

21 Gabrielle wants to see how concession sales changed over the season. She finds the number of hot dogs sold per month. She records these results in a line graph. Were more hot dogs sold in November or February? Explain.

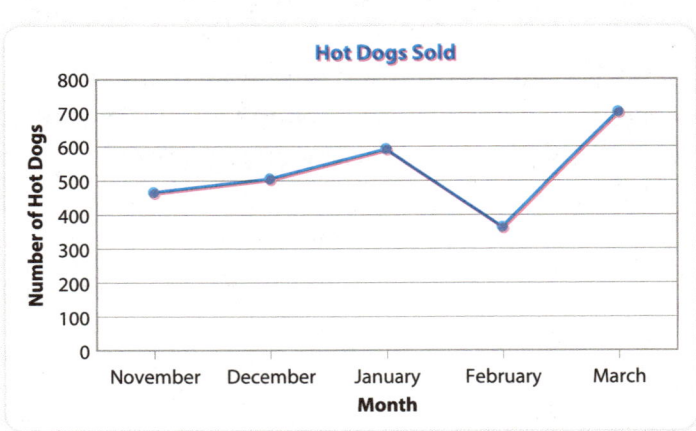

_____

_____

_____

22 Look at problem 21. How did the number of hot dogs sold change throughout the season? _____

_____

_____

Lesson 24 Make Line Graphs and Interpret Data

Lesson 24 👥 **Guided Practice**

Practice: Making Line Graphs and Interpreting Data

Example

Nora's class grew a bean plant from seed. They measured its height every other day and made a line graph. Between which two days did the plant grow the most?

Look how you could use a line graph to answer the question.

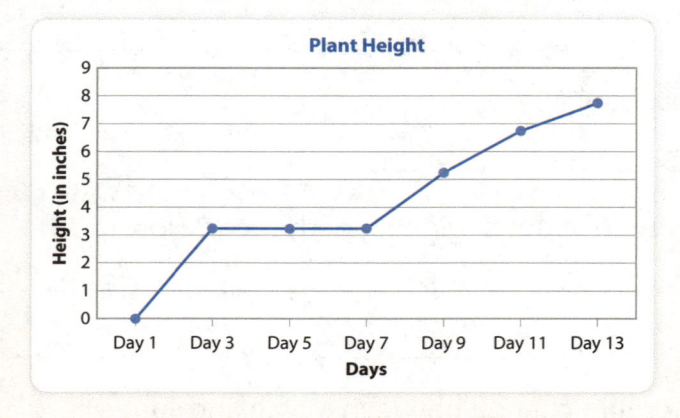

> Look at the *y*-axis to determine the height of the plant for each day.

Plant Height (line graph showing Height in inches vs. Days)

Solution _The plant grew the most between Days 1 and 3._

> **Pair/Share**
> Between which two days did the plant grow the slowest? How do you know?

---

23 A school made a graph showing how many pounds of paper it recycled each month. Between which two months did the amount of recycled paper increase? How do you know?

Pounds of Paper Recycled (line graph showing Pounds Recycled vs. months Sep–May)

> Which direction on a line shows an increase?

> **Pair/Share**
> Between which two months was there the greatest decrease in pounds of paper recycled?

264 Lesson 24 Make Line Graphs and Interpret Data

24 A company measured the number of hours of overtime their employees worked for five days. They found the following data.

Day 1: 9.5 hours; Day 2: 0 hours; Day 3: 6.5 hours; Day 4: 20 hours; Day 5: 4.5 hours.

Make a line graph of the data. Be sure to label each axis, provide a title, and connect the data points with line segments.

What is the greatest number you will need on your graph?

Pair/Share
Is there another way you could have drawn your graph?

25 Which of the following survey questions will produce data that shows a change over time?

  A How many days did it rain last year?
  B How many inches of rain fell each day in April?
  C How much rain falls in each state in July?
  D What months are most likely to have heavy rain?

Cooper chose **A** as the correct answer. How did he get that answer?

_____

_____

_____

Look for signal words and try to predict the type of data.

Pair/Share
What kind of data will each question in the answer choices return?

Lesson 24 👤 Independent Practice

Practice ▶ Making Line Graphs and Interpreting Data

Solve the problems.

1. Masaki tracked the number of hours he spent playing basketball for 5 weeks. He made a line graph of the data.

   Hours Spent Playing Basketball

   Tell whether each sentence is *True* or *False*.

   a. Masaki played the most basketball in Week 5. ☐ True ☐ False

   b. Masaki played basketball for 5 more hours in Week 3 than in Week 1. ☐ True ☐ False

   c. Masaki played fewer hours of basketball in Week 4 than in Week 1. ☐ True ☐ False

   d. The time Masaki played basketball decreased from Week 3 to Week 4. ☐ True ☐ False

2. Look at problem 1. Between which two weeks did Masaki have the greatest increase in the number of hours he played basketball?

   Show your work.

   Answer _____

266 Lesson 24 Make Line Graphs and Interpret Data

3 Which questions will yield data that shows a change over time? Circle all that apply.

  A How many dogs were in the dog park each hour?

  B How much does a dog walker earn per hour?

  C What breeds were at the dog park yesterday?

  D How often do you walk your dog?

  E How much does a golden retriever weigh each year?

4 Decide what type of data each question will produce. Put an X in the correct box.

  a. How often do you ride your bike in one week? ☐ categorical ☐ numerical
  b. Where do you most often ride your bike to? ☐ categorical ☐ numerical
  c. How far do you normally ride your bike? ☐ categorical ☐ numerical

5 Latasha recorded the number of orders placed on her online shop.
Sunday: 44; Monday: 75; Tuesday: 19; Wednesday: 50; Thursday: 10; Friday: 77; Saturday: 80

Part A What type of data did Latasha record?

_____

Part B Make a line graph to display the data.

Part C Between which two days did the number of orders increase the most? Explain.

_____

_____

_____

✓ **Self Check** Go back and see what you can check off on the Self Check on page 235.

Lesson 25 — Introduction
Understand Volume

NC.5.MD.4

Think It Through

What does volume measure?

The **volume** of a three-dimensional figure is the amount of space contained inside the figure. Volume is measured by the number of cubic units that can be packed into a figure.

A cube with edge lengths of 1 unit is called a unit cube. A unit cube has 1 **cubic unit** of volume.

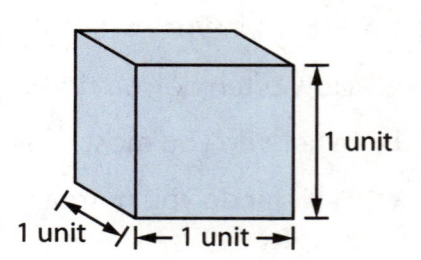

Think How is volume different from area?

Area is the number of *square units* needed to *cover* a *plane figure*. Volume is the number of *cubic units* needed to *fill* a *solid figure*.

A **plane figure** is a two-dimensional figure. To find the area of a plane figure, you need measurements in two dimensions: length and width. One way you can find the area of a plane figure is by covering it in unit squares, without gaps or overlap. A unit square has an area of 1 square unit.

Circle the figures below that have volume.

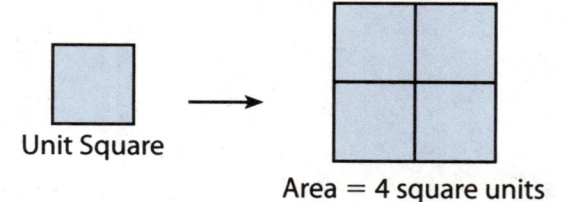

Unit Square → Area = 4 square units

A **solid figure** is a three-dimensional figure. To find the volume of a solid figure, you need measurements in three dimensions: length, width, and height. One way you can find the volume of a solid figure is by packing it with unit cubes, without gaps or overlaps.

Unit Cube → Volume = 4 cubic units

268 Lesson 25 Understand Volume   ©Curriculum Associates, LLC Copying is not permitted.

Think When might you need to measure volume?

Suppose you need to buy a storage box to hold all of your number cubes. All of the cubes are the same size, with edges that are 1 unit long. How would you decide which storage box to buy?

One way to make sure you are buying a big enough box is to use volume. You need a storage box with a volume that will hold all of your cubes.

You know you have 24 number cubes. You can use that number and the volume of each cube to find the volume of the box you need. Then you could make sure that you buy a box big enough to hold all of the cubes.

> I need to count how many cubes are in the whole box, but I can't see all of the cubes. I can see that there are 8 cubes in 1 layer and that there are 3 of the same-size layers.
> 8 + 8 + 8 = 24

← 8 cubes in each layer

Volume = 24 cubic units

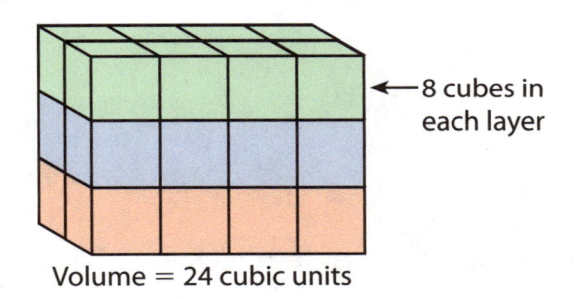

Reflect

1 Explain how 9 square units and 9 cubic units are different.

_____

_____

_____

_____

_____

_____

Lesson 25 *Understand* Volume **269**

Lesson 25 Guided Instruction

Think About: Finding Volume with Unit Cubes

🔍 **Let's Explore the Idea** You can use unit cubes to build or fill a solid figure. The volume of the figure is the number of unit cubes needed to build or fill the solid figure.

2 The measure of each edge length of a unit cube is _____.

3 The volume of each unit cube is _____.

Alexander stacked unit cubes to build the rectangular prism below. Use the rectangular prism to answer problems 4 and 5.

4 There are _____ unit cubes in the bottom layer and there are _____ layers. The figure has _____ unit cubes.

5 The volume of the figure is _____ cubic units.

Now try these two problems. Use the figures below to answer problems 6 and 7. Each figure was built using unit cubes.

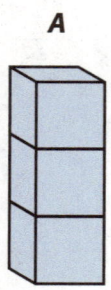

A   B

6 What is the volume of Figure A? _____

7 How many of Figure A does it take to fill or build Figure B? _____

What is the volume of Figure B? _____

270 Lesson 25 Understand Volume

Let's Talk About It
Solve the problems below as a group.

8 Why do you use cubic units instead of square units to find the volumes of solid figures?

_____

_____

9 Candace places blocks with a volume of 1 cubic unit each into her jewelry box, as shown at the right.

How many blocks does it take to fill the jewelry box? _____

What is the volume of Candace's jewelry box? _____

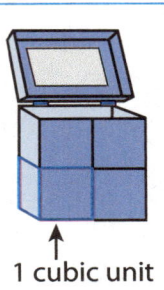

↑ 1 cubic unit

10 Candace has another jewelry box that is like the box in problem 9, but taller. It holds 1 more layer of blocks. What is the volume of this second box?

_____

_____

▶ **Try It Another Way** Work with your group to show the connection between the volume of the solid figure below, the number of layers in the figure, and the number of unit cubes in each layer.

11 How many unit cubes are in each layer? _____

How many layers are in the figure? _____

12 Explain how you can use multiplication to find the volume of the figure. Then find the volume.

_____

_____

Lesson 25 *Understand* Volume **271**

Lesson 25 Guided Practice

Connect: Ideas About Volume with Unit Cubes

Talk through these problems as a class. Then write your answers below.

13 Infer Eli is stacking unit cubes in a box. He partially fills the box, pauses, and says, "The volume of this box is 18 cubic units."

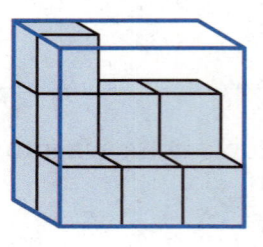

Explain how Eli found the volume of the box. _____
_____
_____

14 Explain Zoe says that a box that is 1 unit wide, 2 units long, and 3 units tall has a greater volume than a box that is 2 units wide, 3 units long, and 1 unit tall. Is she correct? Explain your answer.

_____
_____
_____
_____

15 Compare Each cube in Figures *A* and *B* has a volume of 1 cubic unit. Which figure has less volume, Figure *A* or Figure *B*? Explain your answer.

A   *B*

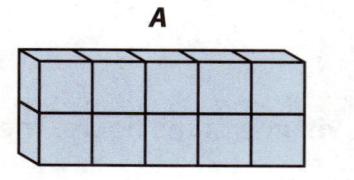

_____
_____
_____

272 Lesson 25 *Understand* Volume

Lesson 25   Independent Practice

Apply ▸ Ideas About Volume with Unit Cubes

16 Put It Together Use what you have learned to complete this task.

> Niles used 16 blocks, each having a volume of 1 cubic unit, to build a rectangular prism.

Part A Draw or build a model to represent the situation.

Part B Look at your model. Describe the number of layers and the number of blocks in each layer. Then describe a different rectangular prism that also has a volume of 16 cubic units.

_____
_____
_____
_____
_____
_____

Lesson 26   Introduction
Find Volume Using Unit Cubes

NC.5.MD.4

Use What You Know

In Lesson 25, you learned that you can fill a solid figure with unit cubes to find its volume. Now take a look at this problem.

> Carl filled the clear box shown below with unit cubes to find its volume.
>
> 2 ft
> 1 ft
> 3 ft
>
> What is the volume of the box?

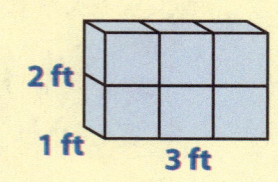

a. How many cubes are in the bottom layer of the box? _____

b. How many cubes are in each layer? _____

c. How many layers of cubes are in the box? _____

d. Should Carl use square units or cubic units to find the volume of the box? Explain.

_____
_____
_____

e. Describe how Carl can find the volume, including what unit he should use.

_____
_____
_____
_____
_____

274   Lesson 26 Find Volume Using Unit Cubes   ©Curriculum Associates, LLC   Copying is not permitted.

Find Out More

Volume is measured in cubic units.

The size of the cubic units used to measure volume can be based on the units used to measure edge lengths. For example, the box on the previous page has edge lengths measured in feet, so its volume is most easily measured in cubic feet.

Each unit cube used to pack the box is 1 foot long, 1 foot wide, and 1 foot high.

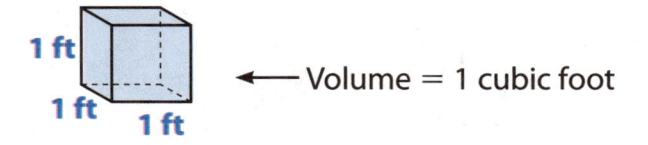

← Volume = 1 cubic foot

One way to find the volume of the box is to count how many cubes fill the box. There are a total of 6 cubes in the box. Each cube has a volume of 1 cubic foot. So, the volume of the box is 6 cubic feet.

Here are two other units you might use to measure volume.

Units of Volume	cubic inch	cubic centimeter
Unit Cube	1 in. × 1 in. × 1 in.	1 cm × 1 cm × 1 cm

Reflect

1 Suppose Carl's box was 4 feet high instead of 2 feet. What would be its volume? Explain your reasoning.

_____

_____

_____

_____

Lesson 26 Find Volume Using Unit Cubes

Lesson 26 Modeled and Guided Instruction

Learn About: Finding the Volume of a Rectangular Prism

Read the problem below. Then explore different ways to find the volume of a rectangular prism.

Abigail used cardboard to build a rectangular prism like the one shown below.

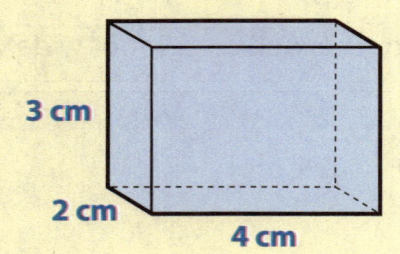

3 cm
2 cm
4 cm

What is the volume of the prism?

Picture It You can find the volume of the prism by filling it with unit cubes and counting the number of cubes.

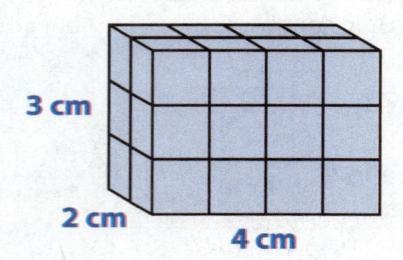

3 cm
2 cm
4 cm

Use cubes that are each 1 cubic centimeter. 24 cubes fill the rectangular prism.

Model It You can also find the volume by counting the number of cubes in one layer and multiplying by the number of layers.

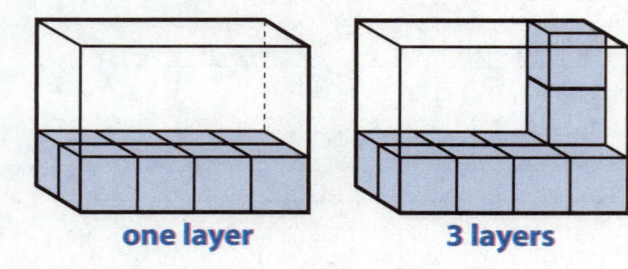

one layer   3 layers

Use cubes that are each 1 cubic centimeter. There are 8 cubes in each layer and 3 layers.

8 cubes × 3 layers = 24 cubes

276 Lesson 26 Find Volume Using Unit Cubes

▶ **Connect It** Now you will solve the problem from the previous page by multiplying.

2 Look at the models on the previous page. How could you find the number of cubes in one layer without counting the cubes? _____

_____

3 Once you know how many cubes are in one layer, what else do you need to know to find the volume? _____

4 Since there are 8 cubes in each layer, and you know there are 3 layers, what multiplication expression can you write to find the volume of the prism? _____

5 What is the volume of Abigail's rectangular prism? _____

6 Explain how you can use multiplication to find the volume of any rectangular prism.

_____

_____

_____

▶ **Try It** Use what you just learned about ways to find the volume of a rectangular prism to solve these problems. Show your work on a separate sheet of paper.

7 What is the volume of the rectangular prism below? _____

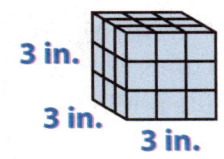

3 in.
3 in.
3 in.

8 Mr. Wong filled the figure below with 1-foot unit cubes to find its volume.

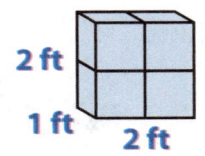

2 ft
1 ft
2 ft

What is the volume of Mr. Wong's figure? _____

Lesson 26 Find Volume Using Unit Cubes **277**

Lesson 26 Guided Practice

Practice: Finding Volume Using Unit Cubes

Study the example below. Then solve problems 9–11.

Example

Pedro has a storage box with a volume of 36 cubic feet. He knows that the box is 4 feet long and 3 feet wide. How high is the box?

Look at how you could show your work using a drawing and multiplication facts.

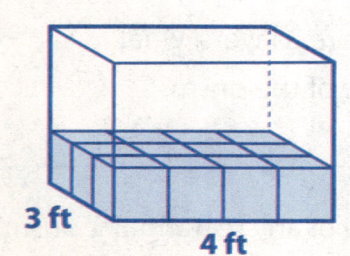

$3 \times 4 = 12$

$12 \times \square = 36$

$\square = 36 \div 12$

$\square = 3$

Solution _3 feet_

The student started by finding the number of cubes in the bottom layer.

Pair/Share
Could you solve this problem another way?

---

9 A box measures 6 centimeters long, 2 centimeters wide, and 4 centimeters high. What is the volume of the box?

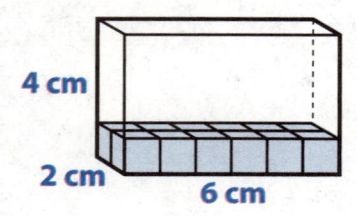

4 cm
2 cm
6 cm

Show your work.

How many layers of cubes will there be in the box?

Pair/Share
Can you use multiplication to solve this problem?

Solution _____

278 Lesson 26 Find Volume Using Unit Cubes

10 Kamala made the figure below using cubes.

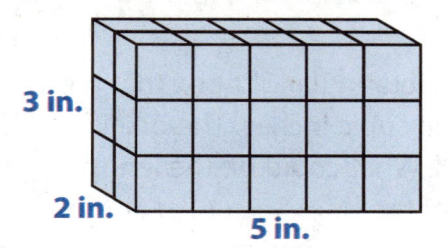

3 in.
2 in.
5 in.

How many cubes are there in each layer?

What is the volume of Kamala's figure?

Show your work.

Pair/Share
How did you decide which method to use to solve the problem?

Solution _____

11 What is the volume of the rectangular prism below?

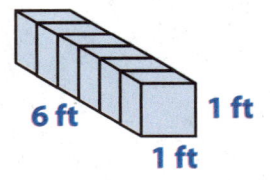

6 ft   1 ft
1 ft

There is more than one way to find the volume of a rectangular prism.

A 6 square feet
B 6 cubic feet
C 8 square feet
D 8 cubic feet

Nam chose **D** as the correct answer. How did he get that answer?

_____

_____

_____

Pair/Share
Does Nam's answer make sense?

Lesson 26 Find Volume Using Unit Cubes

Lesson 26 Independent Practice

Practice ▶ Finding Volume Using Unit Cubes

Solve the problems.

1 How many cubes are in the bottom layer of the rectangular prism below?

- **A** 3
- **B** 6
- **C** 12
- **D** 24

2 Which expressions can be used to find the volume of the rectangular prism below? Circle the letter for all that apply.

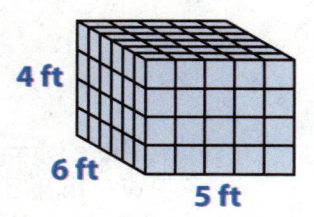

4 ft
6 ft
5 ft

- **A** 30 × 4
- **B** (5 + 6) × 4
- **C** 30 + 30 + 30 + 30
- **D** 5 + 6 + 4
- **E** 30 + 4

3 Flora has a rectangular gift box that has a volume of 24 cubic inches. The box is 2 inches tall. What could be the length and width of the box? Select *Yes* or *No* for each length and width.

a. 11 inches long, 11 inches wide
☐ Yes ☐ No

b. 4 inches long, 3 inches wide
☐ Yes ☐ No

c. 2 inches long, 10 inches wide
☐ Yes ☐ No

d. 6 inches long, 2 inches wide
☐ Yes ☐ No

e. 1 inch long, 12 inches wide
☐ Yes ☐ No

4 Both figures are filled with unit cubes of the same size. Which rectangular prism has the larger volume, figure *A* or figure *B*?

Show your work.

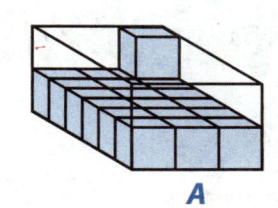

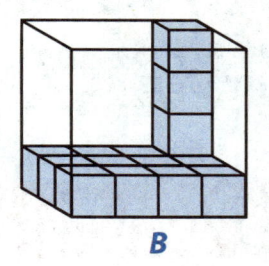

Answer _____ has the larger volume.

5 Mato drew the rectangular prism shown below.

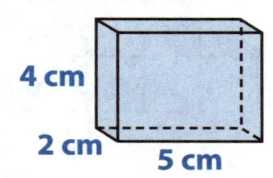

4 cm
2 cm
5 cm

Part A Draw and label a different rectangular prism with the same volume as Mato's prism.

Part B Explain how you know that the volume of your prism is the same as the volume of Mato's prism.

_____

_____

_____

_____

✓ **Self Check**   Go back and see what you can check off on the Self Check on page 235.

©Curriculum Associates, LLC   Copying is not permitted.   Lesson 26 Find Volume Using Unit Cubes

Lesson 27   Introduction
Find Volume Using Formulas

NC.5.MD.5

Use What You Know

In previous lessons, you learned how to find volume by counting unit cubes. In this lesson, you will learn how to find volume using a formula. Take a look at this problem.

> Becky uses 1-inch cubes to create a model for a small paper gift bag she is making. Her model is a rectangular prism. What is the volume of Becky's model?

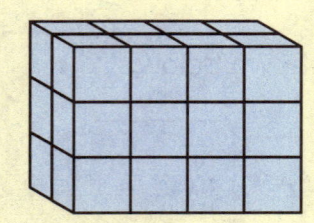

a. The model has a length of _____ inches and a width of _____ inches.

b. How many cubes are in the top layer of the model? _____

c. How can you find the number of cubes in the top layer without counting the cubes?
_____

d. How many layers of cubes does the model have? _____

e. How many cubes are in Becky's model? _____

f. Explain how you can find the volume of Becky's model without counting each cube.
_____
_____
_____
_____

282   Lesson 27 Find Volume Using Formulas

Find Out More

Below is a picture of Becky's gift bag next to her model.

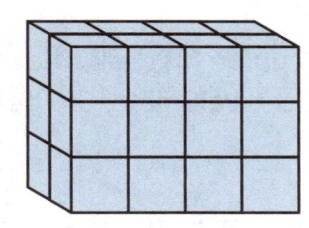

3 in.
2 in.
4 in.

Notice that the base of the bag is a rectangle **4 inches long** and **2 inches wide**. The area of the base is **8 square inches**. The height of the bag is **3 inches**.

Look at the model. There are **8 cubes** in each layer. There are **3 layers** of cubes.

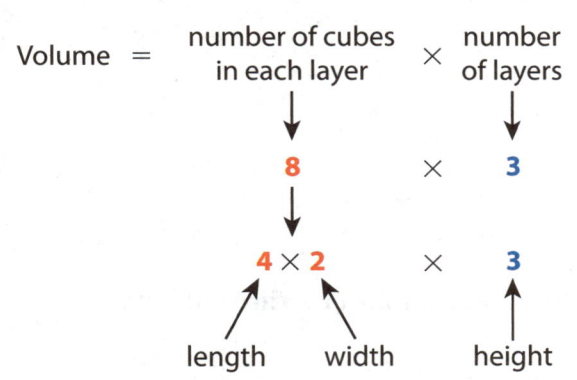

Volume = number of cubes in each layer × number of layers

8 × 3

4 × 2 × 3
length width height

Volume = length × width × height
V = ℓ × w × h

Reflect

1 The area of the base of the gift bag is the product of the length and width of the bag. What is the volume of the bag? Explain how you can use the area of the base of the gift bag to find the volume of the bag.

_____
_____
_____

Lesson 27 Find Volume Using Formulas

Lesson 27   Modeled and Guided Instruction

Learn About: Using a Formula to Find Volume

Read the problem below. Then explore different ways to find the volume of a rectangular prism.

> Gareth has a rectangular pencil cup on his desk. The cup is 3 inches long, 2 inches wide, and 5 inches high. What is the volume of the pencil cup?

▶ **Picture It** You can picture the pencil cup as a rectangular prism made up of 1-inch cubes.

▶ **Model It** You can draw a model of the pencil cup and label its dimensions.

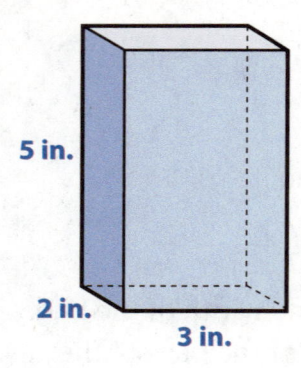

5 in.
2 in.
3 in.

284 Lesson 27 Find Volume Using Formulas   ©Curriculum Associates, LLC   Copying is not permitted.

Connect It Now you will solve the problem from the previous page using the picture and model of the pencil cup.

2 Explain how you can find the volume of the pencil cup using the picture. Find the volume.

_____

_____

_____

3 Explain how the area of the base of the model of the pencil cup in *Model It* relates to the prism in *Picture It*. _____

_____

_____

4 Explain how you can find the volume of the pencil cup using the model.

_____

_____

_____

5 Write two different multiplication equations you can use to find the volume of the pencil cup.

_____

6 Explain how you can use the dimensions of a rectangular prism to find its volume.

_____

_____

Try It Use what you just learned about finding volume to solve these problems. Show your work on a separate sheet of paper.

7 What is the volume of a rectangular jewelry box with a length of 8 centimeters, a width of 5 centimeters, and a height of 4 centimeters? _____

8 How much space is taken up by a book that is 12 inches long, 10 inches wide, and 1 inch tall? _____

Lesson 27 Find Volume Using Formulas

Lesson 27 Guided Practice

Practice: Using the Formula for Volume

Study the example below. Then solve problems 9–11.

Example

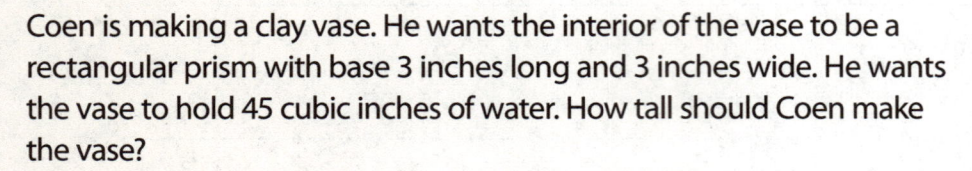

Coen is making a clay vase. He wants the interior of the vase to be a rectangular prism with base 3 inches long and 3 inches wide. He wants the vase to hold 45 cubic inches of water. How tall should Coen make the vase?

Look at how you could use a formula to solve the problem.

$$\text{Volume} = \ell \times w \times h$$
$$45 = 3 \times 3 \times h$$
$$45 = 9 \times h$$
$$45 \div 9 = h$$
$$5 = h$$

Solution ____5 inches____

The student wrote an equation using the formula for volume. The height is the unknown.

Pair/Share
Did you and your partner solve the problem the same way?

9 Evie uses 20 square inches of cardboard for the base of a box. The box has a height of 6 inches. What is the volume of the box?

Show your work.

What does 20 square inches represent?

Pair/Share
What are some possible dimensions of the base of the box?

Solution _____

286 Lesson 27 Find Volume Using Formulas

10 The rectangular prism shown below has a volume of 42 cubic meters. What is the length of the prism?

3 m
2 m
ℓ

What dimensions are labeled on the prism?

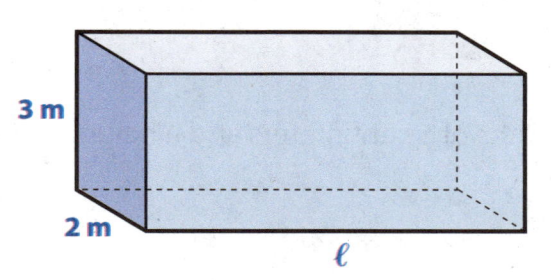

Show your work.

Solution _____

Pair/Share
How could you check your answer?

---

11 A cube is a rectangular prism whose edge lengths are all the same. What is the volume of a cube with an edge length of 2 feet? Circle the letter of the correct answer.

A 4 cubic feet

B 6 cubic feet

C 8 cubic feet

D 12 cubic feet

Danny chose **B** as the correct answer. How did he get that answer?

_____

_____

_____

Each side of a cube is a square.

Pair/Share
Does Danny's answer make sense?

©Curriculum Associates, LLC   Copying is not permitted.   Lesson 27 Find Volume Using Formulas   **287**

Lesson 27 Independent Practice

Practice Using the Formula for Volume

Solve the problems.

1 A rectangular prism has a square base with side lengths of 5 centimeters and a height of 7 centimeters. What is the volume of the prism?

   A 35 cubic centimeters

   B 140 cubic centimeters

   C 175 cubic centimeters

   D 245 cubic centimeters

2 A rectangular prism has a volume of 100 cubic meters. One of the dimensions is 5 meters. Which could be the other two dimensions of the prism? Circle the letter for all that apply.

   A 1 meter, 20 meters

   B 5 meters, 10 meters

   C 10 meters, 10 meters

   D 4 meters, 5 meters

   E 20 meters, 20 meters

3 Henry evenly pours 960 cubic inches of sand into the fish tank shown below. Using the guide on the left side of the tank, shade the front of the tank to show how deep the sand is.

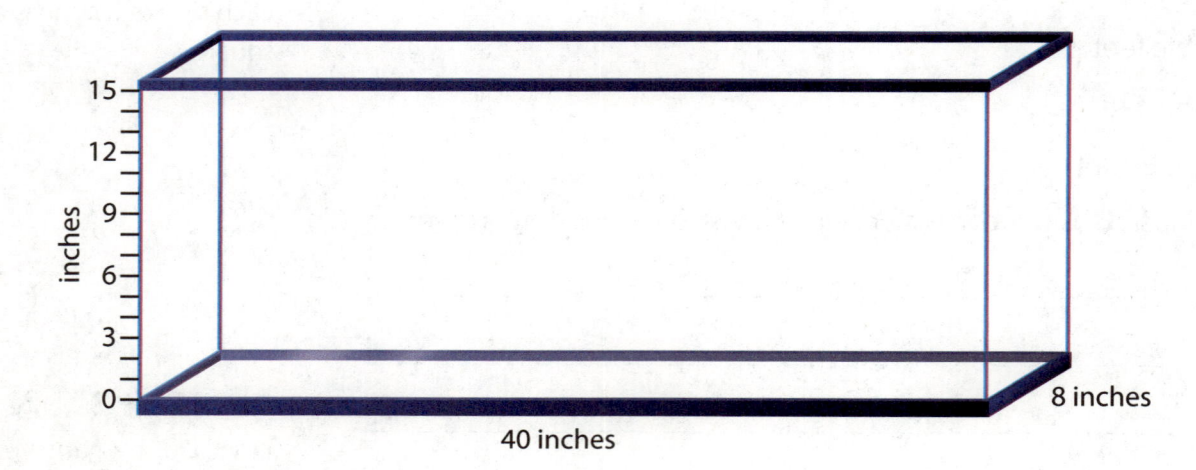

4 A cardboard box has a volume of 60 cubic feet. Give four different sets of measurements that could be the dimensions of the box.

Show your work.

_____ feet × _____ feet × _____ feet

_____ feet × _____ feet × _____ feet

_____ feet × _____ feet × _____ feet

_____ feet × _____ feet × _____ feet

5 The number of guppies that a fish tank can safely hold depends on its volume. A fish tank should have a volume of 576 cubic inches to safely hold 3 guppies. How many guppies can a fish tank that is 24 inches long, 12 inches wide, and 16 inches high safely hold?

Show your work.

Answer _____ guppies

✓ **Self Check** Go back and see what you can check off on the Self Check on page 235.

Lesson 28   Introduction
Find Volume of Composite Figures

Use What You Know

You learned how to use a formula to find the volume of a rectangular prism. In this lesson, you will find the volume of a solid figure made up of more than one rectangular prism. Take a look at this problem.

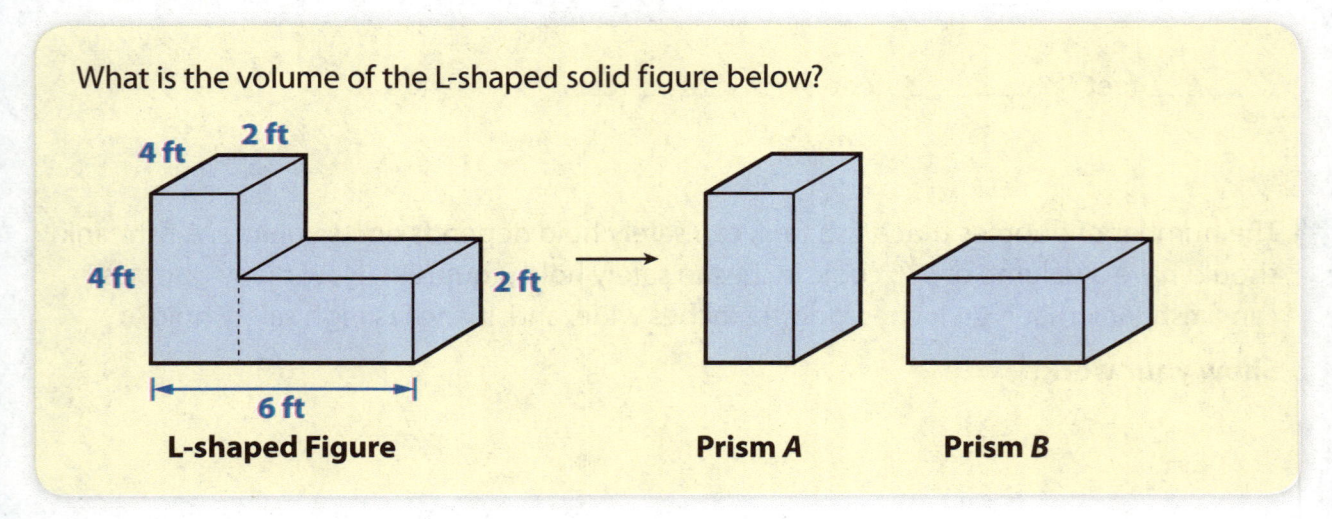

What is the volume of the L-shaped solid figure below?

4 ft, 2 ft, 4 ft, 2 ft, 6 ft

L-shaped Figure → Prism A   Prism B

a. Look at the dashed segment in the L-shaped figure. Imagine slicing the solid figure along the dashed line to make two rectangular prisms.

   What would be the dimensions of Prism A? _____

   What would be the dimensions of Prism B? _____

b. How could you find the volume of each rectangular prism? _____
   _____

c. What is the volume of Prism A? _____

   What is the volume of Prism B? _____

d. How could you find the volume of the L-shaped figure using the two prisms?
   _____

e. What is the volume of the L-shaped solid figure? _____

290   Lesson 28 Find Volume of Composite Figures

Find Out More

Rectangles can be combined in different ways to create different two-dimensional shapes. You can find the areas of these shapes by breaking them apart into the rectangles that make them up. There may be more than one way to break apart these shapes.

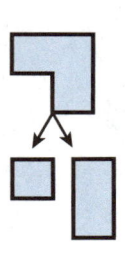

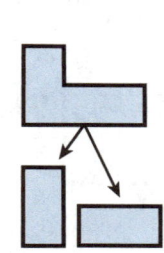

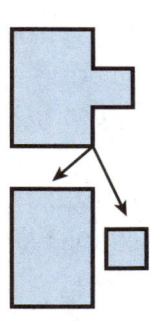

Rectangular prisms can be combined in different ways to make solid figures. You can find the volume of these solid figures by breaking them apart into the prisms that make them up. There may be more than one way to break apart these solid figures.

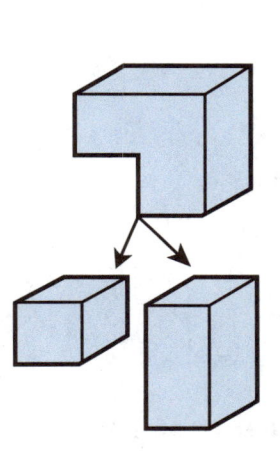

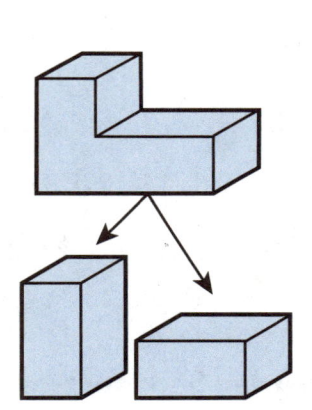

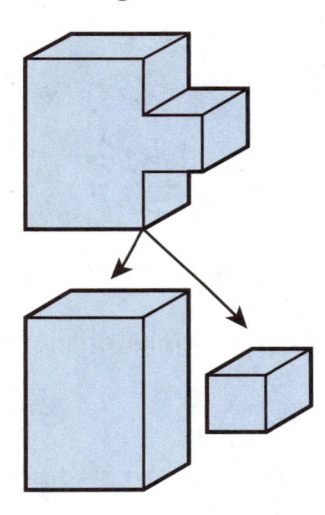

Reflect

1 Circle one of the solid figures above. Explain how you could find its volume.

_____
_____
_____
_____

Lesson 28 Find Volume of Composite Figures

Lesson 28 Modeled and Guided Instruction

Learn About: Breaking Apart Figures to Find Volume

Read the problem. Then explore how to find the volume of a solid figure by breaking it apart into rectangular prisms in different ways.

Bethany has a raised garden bed. The diagram shows its measurements. All of the corners are right angles. If she fills the bed to the top with soil, how many cubic feet of soil will Bethany need?

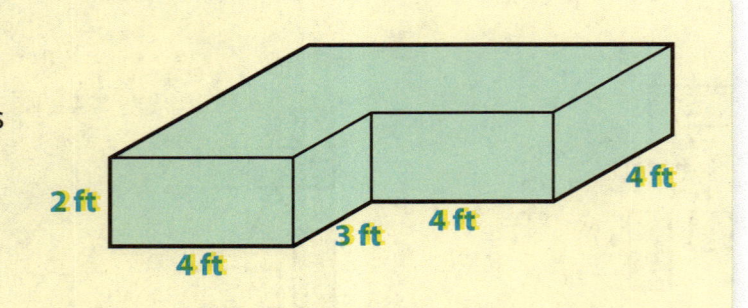

▶ **Model It** You can break the garden bed into two rectangular prisms this way.

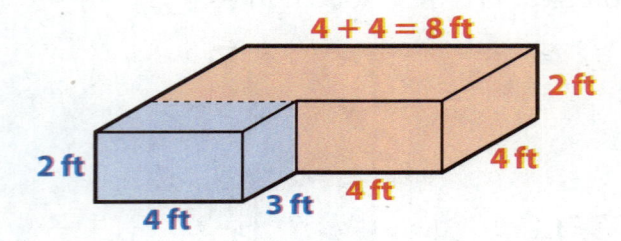

One rectangular prism measures 4 feet × 3 feet × 2 feet.

The other rectangular prism measures 8 feet × 4 feet × 2 feet.

▶ **Model It** You can also break the garden bed into two rectangular prisms in a different way.

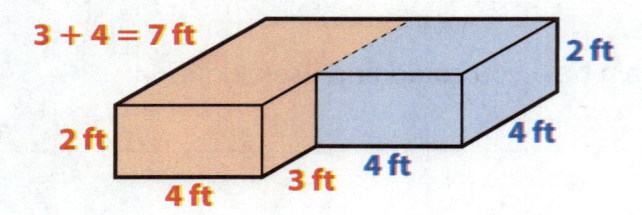

One rectangular prism measures 4 feet × 7 feet × 2 feet.

The other rectangular prism measures 4 feet × 4 feet × 2 feet.

292 Lesson 28 Find Volume of Composite Figures

Connect It Now you will solve the problem from the previous page using the formula for volume of a rectangular prism.

2 Look at the first model. How can you find the volume of each rectangular prism?
_____

3 How can you find the volume of the entire garden bed?
_____

4 What is the volume of the entire garden bed? Show your work. _____

5 Now look at the other model. Show how to find the volume of the garden bed if you break it apart this way.

6 Do you need to break apart a solid figure in a certain way to find its volume? Use the problem from the previous page to explain your reasoning.
_____
_____

Try It Use what you just learned about breaking apart solid figures to solve these problems. Show your work on a separate sheet of paper.

7 The Recreation Center has an L-shaped pool. One part of the pool is 8 meters by 6 meters. The other part is 7 meters by 6 meters. The whole pool is 4 meters deep. What is the volume of the entire pool? _____

8 What is the volume of the solid figure below? _____

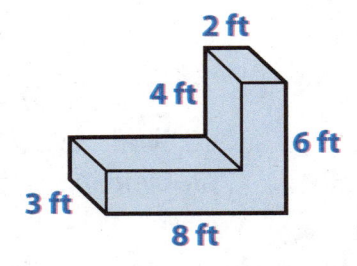

2 ft
4 ft
6 ft
3 ft
8 ft

Lesson 28 Find Volume of Composite Figures **293**

Lesson 28 Guided Practice

Practice: Finding the Volume of Composite Figures

Study the example below. Then solve problems 9–11.

Example

Adam drew a model of a flowerpot. All of its edges meet at right angles. What is the volume of the flowerpot?

Look at how you could break apart the figure into two rectangular prisms.

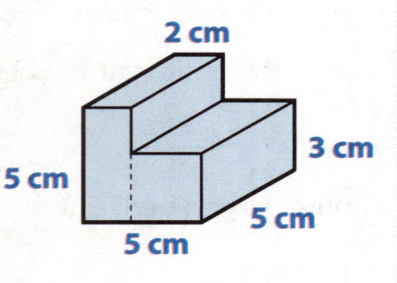

$2 \times 5 \times 5 = 50$ $3 \times 5 \times 3 = 45$

$50 + 45 = 95$

Solution 95 cubic centimeters

The student drew a dashed segment to show how to break apart the solid figure.

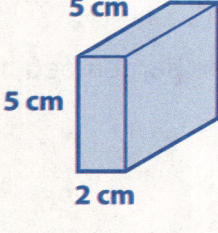

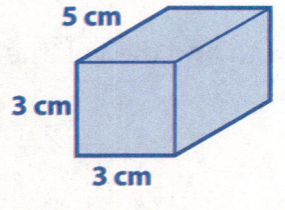

Pair/Share
Could you break apart the solid figure in a different way?

---

9 Tia made a small two-layer cake. Each layer is a rectangular prism that is 2 inches high. The base of the bottom layer is 8 square inches. The base of the top layer is also 8 square inches. What is the total volume of the cake?

Show your work.

Would it help to draw a diagram to represent the problem?

Pair/Share
Can you solve the problem in a different way?

Solution _____

294   Lesson 28 Find Volume of Composite Figures

10 The diagram shows the dimensions of a cement walkway, where all of the sides meet at right angles. What is the total volume of cement needed to make the walkway?

After I break apart the figure, how will I find the missing measurements?

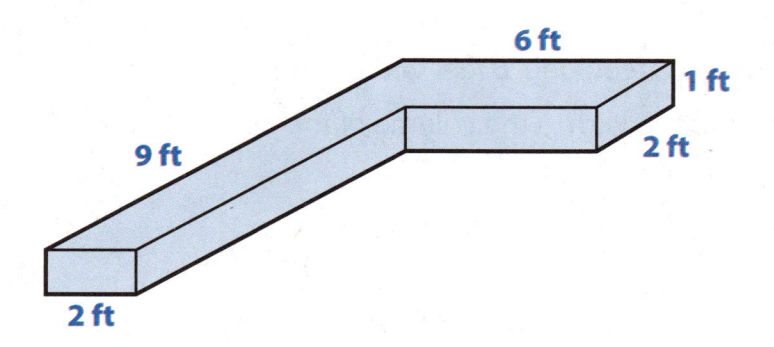

Show your work.

Pair/Share
How did you and your partner decide how to break apart the solid figure?

Solution _____

---

11 A candy store is building a countertop display with two rectangular prisms. The base of each prism is a square with a edge length of 5 inches. One prism is 4 inches tall and the other is twice as tall. The two prisms are stacked so that the pieces do not overlap. What is the total volume of the display?

A 12 cubic inches

B 20 cubic inches

C 60 cubic inches

D 300 cubic inches

Kali chose **A** as the correct answer. How did she get that answer?

What is the total height of the display?

Pair/Share
Does Kali's answer make sense?

Lesson 28 Find Volume of Composite Figures **295**

Lesson 28 Independent Practice

Practice Finding the Volume of Composite Figures

Solve the problems.

1 A pool shaped like a rectangular prism is 8 feet long, 6 feet wide, and 8 feet deep. Attached to the side of the pool is a hot tub with $\frac{1}{4}$ the volume of the pool. What is the total volume of the pool and hot tub?

A 96 cubic feet

B 60 cubic feet

C 384 cubic feet

D 480 cubic feet

2 The diagram below shows the measurements of a mold used to make sandcastles. Which expression can be used to find the volume of the mold, in cubic inches? Circle the letter for all that apply.

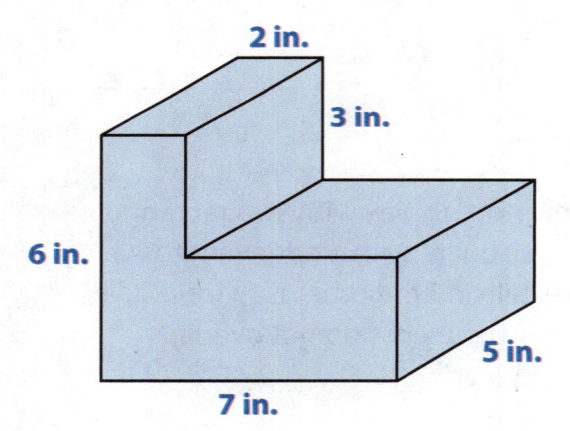

A $(2 \times 5 \times 3) + (5 \times 5 \times 3)$

B $(2 \times 5 \times 3) + (7 \times 5 \times 3)$

C $(2 \times 5 \times 6) + (5 \times 5 \times 3)$

D $(2 \times 5 \times 6) + (7 \times 5 \times 3)$

E $(2 \times 5 \times 3) + (7 \times 5 \times 6)$

3 The diagram below shows the dimensions of two identical rectangular prisms joined together.

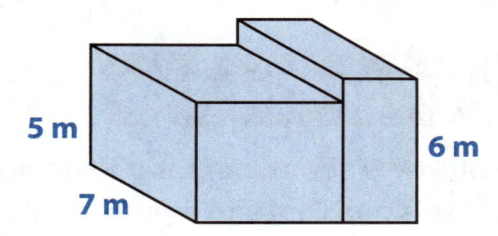

5 m
6 m
7 m

What is the combined volume of the two prisms? _____ cubic meters

4 Rami designed a small pond for a restaurant. The diagram below shows the measurements of the pond. How many cubic feet of water are needed to fill the pond?

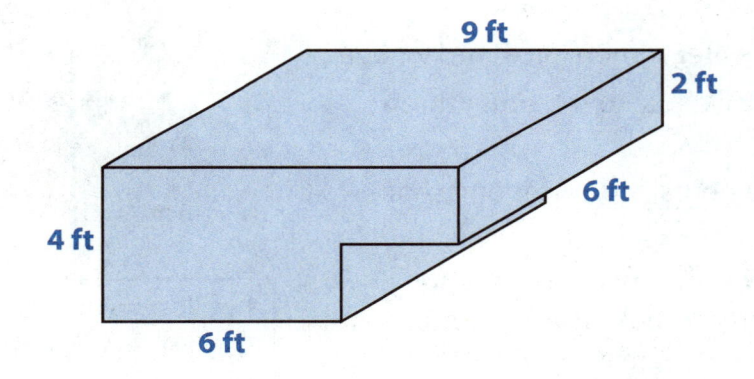

9 ft
2 ft
6 ft
4 ft
6 ft

Show your work.

Answer _____ cubic feet

✓ **Self Check** Go back and see what you can check off on the Self Check on page 235.

©Curriculum Associates, LLC   Copying is not permitted.   Lesson 28 Find Volume of Composite Figures   **297**

Unit 4 MATH IN ACTION

Introduction
Work with Measurement and Data

SMP1 Make sense of problems and persevere in solving them.

Study an Example Problem and Solution

In this lesson, you will use what you know about measurement and data to solve real-world problems. Look at this problem and one solution.

Salad Dressing

Sweet T needs to make salad dressing for an event. He knows that 2 gallons is not enough and 3 gallons is too much. Sweet T finds 2 quarts of vinegar and 2 gallons of oil in the cabinet. Read the salad dressing recipe.

Salad Dressing Recipe
- Mix equal parts of water, lemon juice, and vinegar.
- Add an amount of oil equal to the other three ingredients combined.
- Mix in your favorite herbs and spices and shake.

Sweet T wants to know how much water, lemon juice, vinegar, and oil he needs to make the dressing. He also wants to know how much oil and vinegar will be left over.

1 gallon = 16 cups
1 gallon = 4 quarts
1 quart = 4 cups

Read the sample solution on the next page. Then look at the checklist below. Find and mark parts of the solution that match the checklist.

Problem-Solving Checklist

☐ Tell what is known.
☐ Tell what the problem is asking.
☐ Show all your work.
☐ Show that the solution works.

a. **Circle** something that is known.
b. **Underline** something that you need to find.
c. **Draw a box around** what you do to solve the problem.
d. **Put a checkmark** next to the part that shows the solution works.

Sweet T's Solution

> Hi, I'm Sweet T. Here's how I solved this problem.

▷ **I know** the four different ingredients and how the amounts compare.

> 1 gallon = 16 cups
> 1 gallon = 4 quarts
> 1 quart = 4 cups

▷ **I have to decide** what units to use to find the amount of each ingredient and the amount left over. The information in the problem has both quarts and gallons. I think it will be easier to convert all amounts to cups.

There are 16 cups in a gallon.
2 gallons = 2 × 16, or 32 cups
3 gallons = 3 × 16, or 48 cups

There are 4 cups in a quart.
2 quarts = 2 × 4, or 8 cups

> I can also use quarts, but I might end up using cups to find leftover amounts anyway.

▷ **Now I know** that the total amount is greater than 32 cups and less than 48 cups.

▷ **One way to think about** the amounts is that half of the dressing is oil and half is a combination of water, lemon juice, and vinegar. I'll look for half of a number between 32 and 48.

▷ **I'll use 36** for the total.
$\frac{1}{2} \times 36 = 18$ so I need 18 cups of oil.
Then divide by 3 to find the other amounts.
18 ÷ 3 = 6 so I need 6 cups of each of the other ingredients.

> I multiplied 36 by $\frac{1}{2}$ but I also could have divided by 2.

▷ **I can check by adding.**
18 cups + 6 cups + 6 cups + 6 cups = 36 cups
36 > 32 and 36 < 48

▷ **Now I can write** the amount of each ingredient and find the leftovers.

Amount for dressing
18 c oil
6 c vinegar
6 c water
6 c lemon juice

Amount leftover
32 c − 18 c = 14 c leftover oil
8 c − 6 c = 2 c leftover vinegar

> I used the abbreviation for cups since I had to write it so many times!

Unit 4 Math in Action Modeled and Guided Instruction

Try Another Approach

There are many ways to solve problems. Think about how you might solve the Salad Dressing problem in a different way.

Salad Dressing

Sweet T needs to make salad dressing for an event. He knows that 2 gallons is not enough and 3 gallons is too much. Sweet T finds 2 quarts of vinegar and 2 gallons of oil in the cabinet. Read the salad dressing recipe.

Salad Dressing Recipe
- Mix equal parts of water, lemon juice, and vinegar.
- Add an amount of oil equal to the other three ingredients combined.
- Mix in your favorite herbs and spices and shake.

Sweet T wants to know how much water, lemon juice, vinegar, and oil he needs to make the dressing. He also wants to know how much oil and vinegar will be left over.

1 gallon = 16 cups
1 gallon = 4 quarts
1 quart = 4 cups

▶ **Plan It** Answer these questions to help you start thinking about a plan.

A. How could you convert the ingredients to quarts to find a solution?
_____
_____

B. What are some total amounts of dressing that you could choose? Explain.
_____
_____

▶ **Solve It** Find a different solution for the Salad Dressing problem. Show all your work on a separate sheet of paper.

You may want to use the problem-solving tips to get started.

Problem-Solving Tips

- **Tools** You might want to use . . .
  - a table.
  - a bar model.

- **Word Bank**

cup	gallon	subtract
pint	convert	divide
quart	add	

- **Sentence Starters**
  - One quart is equal _____
  - The total amount _____

Problem-Solving Checklist
Make sure that you . . .
☐ tell what you know.
☐ tell what you need to do.
☐ show all your work.
☐ show that the solution works.

▶ **Reflect**

Use Mathematical Practices As you work through the problem, discuss these questions with a partner.

- **Persevere** How can you use your answers to the *Plan It* questions to help choose a solution path?
- **Be Precise** Why should you label your numbers with units as you work through the solution?

Unit 4 Math in Action — Guided Practice

Discuss: Models and Strategies

Read the problem. Write a solution on a separate sheet of paper. Remember, there can be lots of ways to solve a problem!

Goldfish Pool

Sweet T is helping design a small goldfish pool to be built in front of his friends' bakery store.

The pool will be shaped like the diagram shown below. The two ends are the same depth. The middle section is deeper.

Pool Plans

- The total length is between 6 and 9 feet.
- The width is between 6 and 8 feet.
- The deepest part of the pool is no more than 4 feet deep.
- The total volume is 250 cubic feet or less.

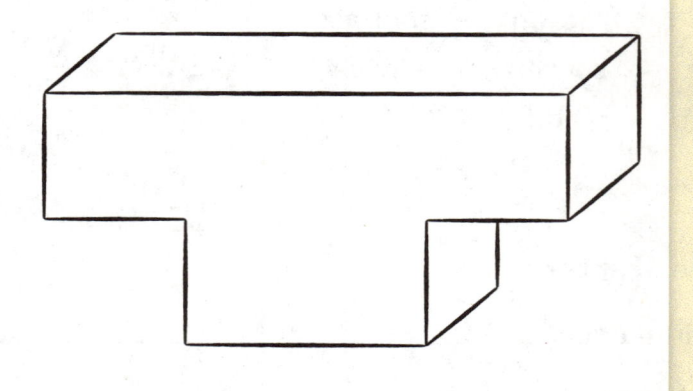

What should the dimensions of the pool be?
What will the volume of the pool be?

▶ **Plan It and Solve It** Find a solution to the Goldfish Pool problem.

Use the Goldfish Pool Activity Sheet.
- Fill in all the missing dimensions.
- Verify that the lengths, widths, and depths you chose meet all of the requirements in the pool plans.
- Find the volume of the pool.

Problem-Solving Tips

- **Questions**
  - How deep will the shallow section be? The deep section?
  - How can you divide the solid figure into rectangular prisms?

- **Word Bank**

rectangular prism	length	volume
solid figure	width	cubic feet

Problem-Solving Checklist
Make sure that you . . .
☐ tell what you know.
☐ tell what you need to do.
☐ show all your work.
☐ show that the solution works.

▶ **Reflect**
Use Mathematical Practices As you work through the problem, discuss these questions with a partner.

- **Use a Model** How can the model help you find appropriate dimensions?
- **Be Precise** What measurement units will you use when you work out the solution? Explain.

Unit 4 Math in Action   **Independent Practice**

Persevere ▶ On Your Own

Read the problems. Write a solution on a separate sheet of paper. Remember, there are many different ways to solve a problem.

Backyard Barbecue

Sweet T is planning a barbecue for 50 people. There will be 2 different kinds of protein and 3 side dishes on the menu. Here are his choices, including amounts to estimate per person.

Protein
- Choose from ground beef, chicken, steak, or salmon.
- Estimate 6 to 8 ounces per person.

Sides
- Baked beans: 2 to 3 ounces per person
- Coleslaw: 3 to 4 ounces per person
- Potato salad: 4 to 5 ounces per person
- Grilled vegetables: 3 to 4 ounces per person
- Rice: 1 to 2 ounces per person

1 pound = 16 ounces

What food and how much of each should Sweet T make?

▶ Solve It Suggest a menu for Sweet T.
- Choose the items for the menu.
- Tell how many pounds and ounces are needed for each dish. Explain.

▶ Reflect

Use Mathematical Practices After you complete the task, choose one of these questions to discuss with a partner.

- **Make Sense of Problems** What was the first step you took to find your solution? Why?
- **Reason Mathematically** What operations did you use to find your solution? Explain why.

Layered Dessert

Sweet T is making his favorite layered dessert. Read his notes.

Layered Dessert Notes
- Cut brownies, marshmallows, and cake into cubes.
- First layer is brownie cubes.
- Second layer is marshmallow cubes.
- Third layer is cake cubes.
- Use more than 3 layers.
- You choose the thickness of each layer.
- Repeat layers as many times as you want to fill the bowl.

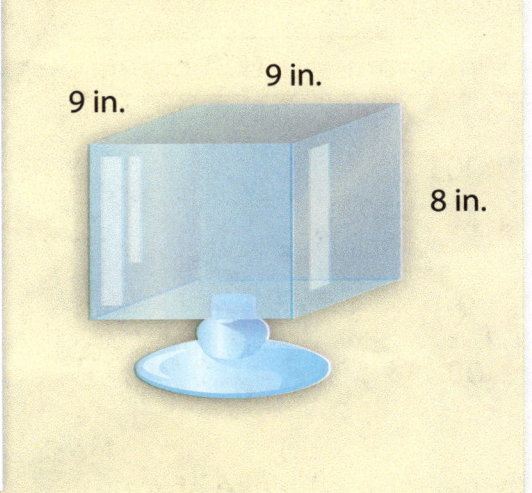

9 in.
9 in.
8 in.

The picture above shows the container Sweet T uses to make the dessert. How many layers should Sweet T use? How thick should each layer be?

▶ **Solve It** Help Sweet T make a plan.
- Tell which item is in each layer.
- Give the length, width, and volume of each layer.
- Find the total volume of the completed dessert.

▶ **Reflect**

Use Mathematical Practices After you complete the task, choose one of these questions to discuss with a partner.

- **Use a Model** What models did you use and how did they help you find a solution?
- **Persevere** What did you do to get through any difficult parts of the solution?

©Curriculum Associates, LLC   Copying is not permitted.   Unit 4 Math in Action Work with Measurement and Data   **305**

Unit 4 Assessment
Interim Assessment

Solve the problems.

1. Regan wants to convert 2 kilograms to grams. What number does she need to multiply by 2 to find the number of grams?

   | 1 kilogram = 1,000 grams |

   A   0.001

   B   0.1

   C   100

   D   1,000

2. The owner of a cupcake store tracked the number of cupcakes she sold daily for 1 week. The owner made a line graph of the data.

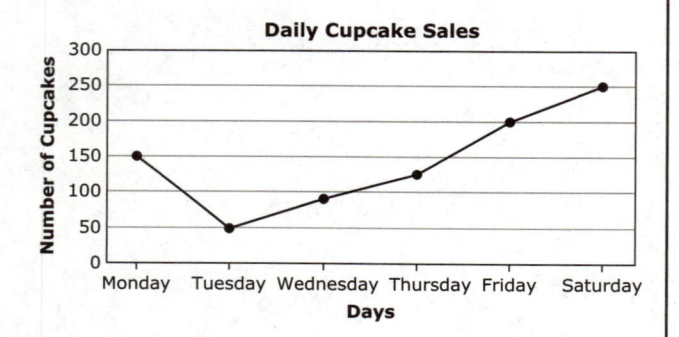

   Daily Cupcake Sales

   Between which two days did the number of cupcakes sold increase the most?

   A   Monday and Tuesday

   B   Wednesday and Thursday

   C   Thursday and Friday

   D   Friday and Saturday

3. The solid figure below is made up of two rectangular prisms.

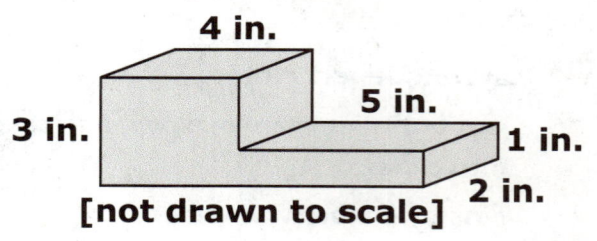

   4 in.
   5 in.
   3 in.
   1 in.
   2 in.
   [not drawn to scale]

   What is the volume of the solid figure?

   A   24 cubic inches

   B   34 cubic inches

   C   44 cubic inches

   D   54 cubic inches

4. An aquarium is shaped like a rectangular prism. It is 24 inches long and 10 inches wide. Its volume is 2,880 cubic inches. How high is the aquarium?

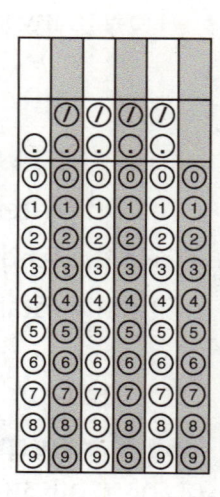

5. How many feet and inches are equivalent to 86 inches?

   | 1 foot = 12 inches |

   A   7 feet, 0 inches

   B   7 feet, 2 inches

   C   8 feet, 0 inches

   D   8 feet, 2 inches

306   Unit 4 Interim Assessment

6. Last week Jaylen ran 2.4 kilometers. How many meters did he run?

   1 kilometer = 1,000 meters

   A   0.024 meter

   B   24 meters

   C   240 meters

   D   2,400 meters

7. During the last three weeks Skylar ran a total of 9 hours and 13 minutes. How many total minutes did Skylar run?

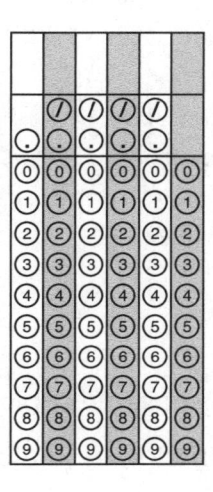

   60 minutes = 1 hour

8. The figure below is a rectangular prism.

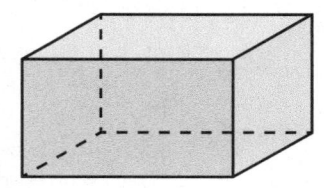

   Which statement explains how you can use the area of the base to find the volume of a rectangular prism?

   A   Multiply the area of the base times the height of the rectangular prism.

   B   Subtract the height of the rectangular prism from the area of the base.

   C   Divide the height of the rectangular prism by the area of the base.

   D   Add the area of the base to the height of the rectangular prism.

Unit 4 Interim Assessment continued

Performance Task

Answer the questions and show all your work on separate paper.

Jack wants to make a terrarium for the science fair. The first steps of the instructions are shown below.

How To Make a Terrarium

Step 1 Fill the bottom of the tank with a 2-inch layer of small rocks.

Step 2 Add a layer of soil. There should be 5 inches from the top of the soil to the top of the tank.

Checklist
Did you . . .
☐ draw a diagram?
☐ convert the units?
☐ use a formula?

Jack already has one cubic foot (1,728 cubic inches) of soil and $\frac{1}{3}$ cubic foot (576 cubic inches) of small rocks to use. Now he has to choose a tank. Which of the tanks below can he use without having to buy more rocks and soil?

| 1 foot = 12 inches |

Tank A: $1\frac{1}{2}$ feet long, 15 inches wide, 15 inches tall

Tank B: 2 feet long, 1 foot wide, 1 foot tall

Tank C: 21 inches long, 18 inches wide, 10 inches tall

▶ **Reflect**

Use Mathematical Practices After you complete the task, choose one of the following questions to answer.

- **Be Precise** How did you decide which units to use to calculate the amount of rocks and soil needed for each tank?

- **Model** How does drawing and labeling a diagram help you solve the problem?

Unit 5
Geometry

Let's learn about the coordinate plane.

Real-World Connection Knowing how to tell where a place is located is important. You might make arrangements to meet someone at a certain location or you might need to give someone directions to a location.

For example, you might need to find the subway station at the corner of 23rd Street and Broadway in New York City. Describing locations is even important in games! In chess, you can move your pawn from c2 to c4 on the chess board. Other players know the direction you moved based on the description. Both these situations use coordinates.

In This Unit You will learn to locate points on a coordinate plane and use what you know about coordinates to solve problems. You will also study quadrilaterals and sort them into different categories according to their side lengths, angles, and other properties.

✓ Self Check

Before starting this unit, check off the skills you know below. As you complete each lesson, see how many more skills you can check off!

I can:	Before this unit	After this unit
graph points on a coordinate plane.	☐	☐
find the distance between two points on a coordinate plane.	☐	☐
graph quantities that represent real-world situations on a coordinate plane and use the graph to solve a problem.	☐	☐
classify quadrilaterals based on their properties, for example: a square is also a rhombus or rectangle, but not all rhombuses and rectangles are squares.	☐	☐

Lesson 29 — Introduction
Understand the Coordinate Plane

NC.5.G.1

Think It Through

What is the coordinate plane?

In the past, you have used a number line to represent a single quantity. A number line can be vertical or horizontal.

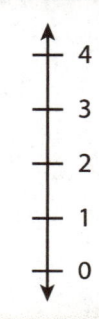

When a horizontal number line and a vertical number line are lined up so that the 0s meet, a **coordinate plane** is formed.

Vertical Number Line

Horizontal Number Line

Coordinate Plane

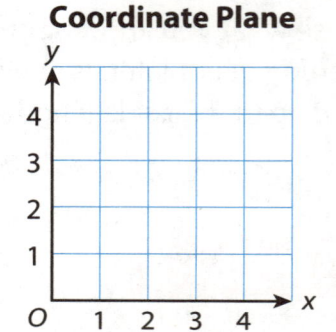

Think How do we distinguish between the two number lines?

A coordinate plane is made up of two perpendicular number lines. The number lines meet at the zeros. That point is labeled with the letter O for origin.

Coordinate Plane

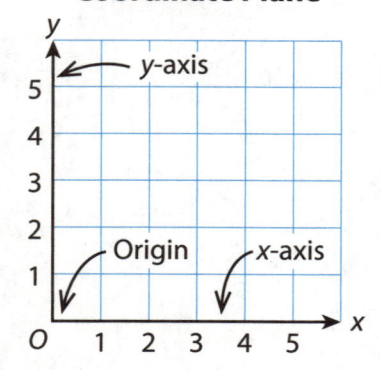

- y-axis
- Origin
- x-axis

The horizontal number line is called the **x-axis**. The vertical number line is called the **y-axis**. The point where the x- and y-axes meet is the **origin**.

Circle all of the numbers on the x-axis. **Box** all of the numbers on the y-axis.

310 Lesson 29 Understand the Coordinate Plane

> **Think** How can you locate points on the coordinate plane?

An **ordered pair** is a pair of numbers that describes the location of a point in the coordinate plane. The first number is called the **x-coordinate**. The second number is called the **y-coordinate**.

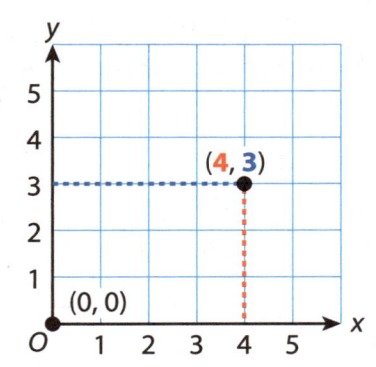

Ordered Pair
(x, y)
x-coordinate   y-coordinate

An ordered pair is a pair of coordinates that always appear in the same order: first x, then y.

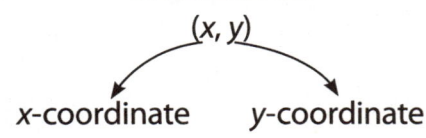

The *x*-coordinate tells the distance from the origin along the *x*-axis. The *y*-coordinate tells the distance from the origin along the *y*-axis.

The origin is the point where the *x*-axis and the *y*-axis meet. Since the origin is 0 units along the *x*-axis and 0 units along the *y*-axis, the location is (0, 0).

Term	Definition
coordinate plane	a space formed by two perpendicular number lines called axes
x-axis	the horizontal number line in the coordinate plane
y-axis	the vertical number line in the coordinate plane
origin	the point, (0, 0), where the *x*-axis and *y*-axis intersect
ordered pair	a pair of numbers that describes the location of a point in the coordinate plane
x-coordinate	the first number in an ordered pair
y-coordinate	the second number in an ordered pair

▶ **Reflect**

1 Think about how you have heard the word *origin* used outside of math. Why do you think the point (0, 0) is called the *origin*?

_____

_____

_____

Lesson 29 Understand the Coordinate Plane **311**

Lesson 29 Guided Instruction

Think About ▸ Points on a Coordinate Plane

🔍 **Let's Explore the Idea** Ordered pairs represent the location of points on the coordinate plane. The ordered pair for the origin is (0, 0).

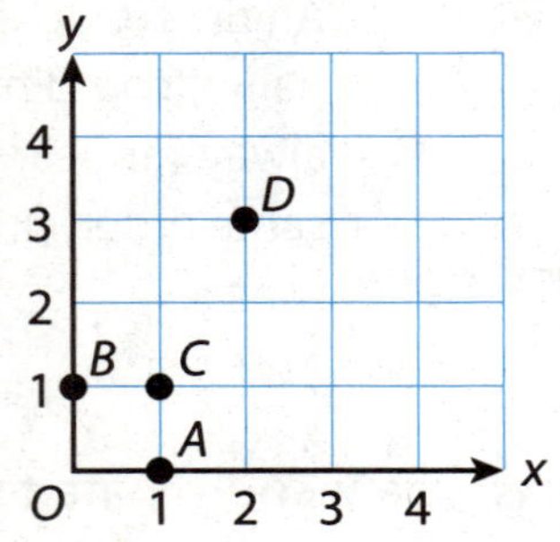

2 Locate point *D* in the coordinate plane above. Start at the origin. How many units do you move along the *x*-axis to be lined up under point *D*? _____ Write this as the first number in the ordered pair at the right.

Ordered Pair for Point *D*

(x, y) ⟶ (_____ , _____)

3 The second number in an ordered pair is the *y*-coordinate. Start at the origin. How many units do you move along the *y*-axis to be lined up across from point *D*? _____ Write this as the second number in the ordered pair above.

4 Locate point *A* in the coordinate plane above. Start at the origin. How many units do you move along the *x*-axis to be on point *A*? _____
Start at the origin. How many units do you move along the *y*-axis to be lined up across from point *A*? _____

The ordered pair for point *A* is (_____ , _____).

5 Locate point *B* in the coordinate plane above. Identify the ordered pair for point *B*.
(_____ , _____)

6 Locate point *C* in the coordinate plane above. Identify the ordered pair for point *C*.
(_____ , _____)

7 How did you determine the coordinates for points *C* and *D*? _____

_____

_____

312 Lesson 29 *Understand* the Coordinate Plane   ©Curriculum Associates, LLC Copying is not permitted.

Let's Talk About It
Solve the problems below as a group.

Use the coordinate plane from the previous page. Now you will describe the points as movements from the origin.

8 Place your finger on the origin. Move right and then move up to point *D*. Use words to describe how far to the right and how far up point *D* is from (0, 0). _____

_____

How is the location of point *D* related to the ordered pair (2, 3)? _____

_____

9 Start at (0, 0). Move to point *A*. How is your move related to the ordered pair (1, 0)?

_____

_____

10 Describe how you move from the origin to the point (0, 1). Move right _____ unit(s) and move _____ _____ unit(s).

What letter represents this point on the coordinate plane? _____

11 The ordered pair (1, 1) tells you to start at the origin and move right _____ unit(s) and up _____ unit(s). What letter names this point on the coordinate plane? _____

12 Plot a point at (3, 2) on the coordinate plane. Label this point "*E*."

Try It Another Way
Work with your group to show the coordinates for each point on the coordinate plane on the previous page.

13 Complete the table for each point on the coordinate plane.

Point	A	B	C	D	E	F
x	1				3	4
y	0				2	0

14 The table gives the location of point *F*. Plot and label point *F* in the coordinate plane.

Lesson 29 *Understand the Coordinate Plane*

Lesson 29 **Guided Practice**

Connect **Ideas About Points on a Coordinate Plane**

Talk through these problems as a class. Then write your answers below.

15 Show Look at the table below. Plot and label points *M* and *N* on the coordinate plane below. Then write ordered pairs for points *M* and *N* and describe how to move from (0, 0) to each point.

Point	x	y
M	1	4
N	5	2

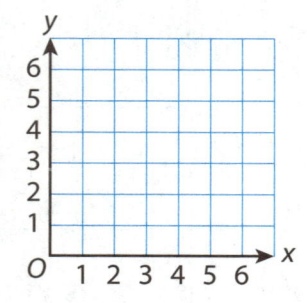

_____

_____

16 Analyze Irvin wrote the ordered pair (4, 3) for the location of point *J* in the coordinate plane at the right. Explain why Irvin's ordered pair is incorrect for point *J*.

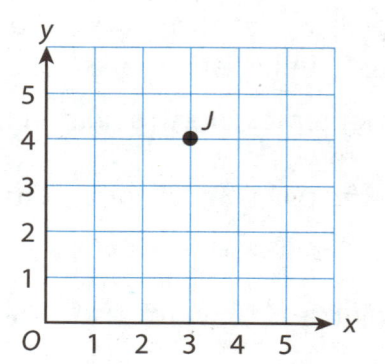

_____

_____

_____

17 Create Choose 3 points on the coordinate plane at the right and draw a triangle. Label the points with letters.

Below, write the letters and ordered pairs you used to draw your triangle.

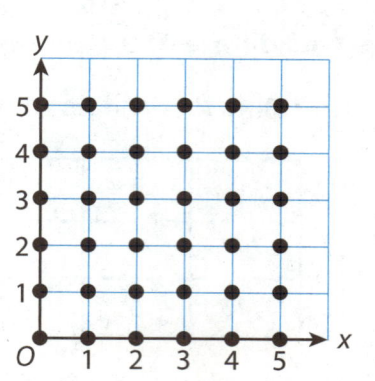

_____

_____

_____

314 Lesson 29 *Understand* the Coordinate Plane

Lesson 29 — Independent Practice

Apply: Ideas About Points on a Coordinate Plane

18 Put It Together Use what you have learned to complete this task.

Part A Use the coordinate plane below to complete the table.

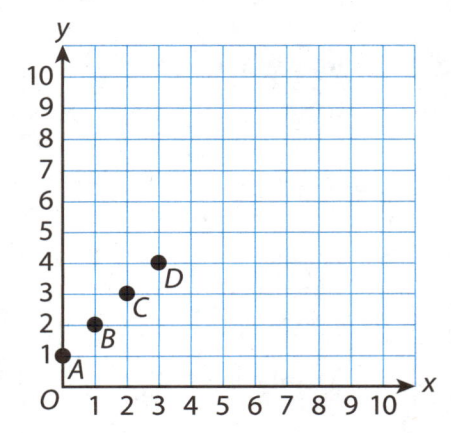

Point	x	y
A		
B		
C		
D		

Part B Identify a pattern you see formed by the points in the coordinate plane above. Then, explain the pattern using the points in the table above.

_____
_____
_____
_____
_____
_____

Part C Plot 3 other points that follow this pattern. Label them with letters and write the ordered pair for each point below.

_____
_____
_____

©Curriculum Associates, LLC   Copying is not permitted.

Lesson 29 *Understand* the Coordinate Plane   **315**

Lesson 30 Introduction
Graph Points in the Coordinate Plane

NC.5.G.1

Use What You Know

In Lesson 29, you learned about the coordinate plane. Now you will use the coordinate plane to solve problems. Take a look at this problem.

> The coordinate plane at the right shows the layout at a country fair. Some of the booths are labeled. Meg is at Ring Toss. She wants to go to Face Painting. Suppose Meg can only walk along the grid lines. Describe two different paths she can take.

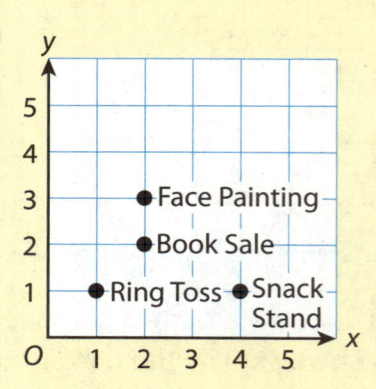

a. What ordered pair represents Ring Toss? _____

b. What ordered pair represents Face Painting? _____

c. Describe one way to move from Ring Toss to Face Painting by first moving to the right.

_____

_____

_____

d. Now describe a different path to get from Ring Toss to Face Painting.

_____

_____

_____

_____

316 Lesson 30 Graph Points in the Coordinate Plane

▶▶ Find Out More

On the previous page you used the coordinate plane to help you describe how to move between points. But you can do this without using the coordinate plane. You can use ordered pairs to describe how to move between points.

Use the ordered pairs (1, 1) and (2, 3) to describe how to move between the points. To describe the horizontal move, look at the change in the *x*-coordinates.

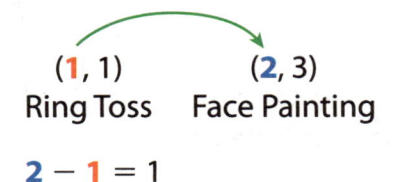

(**1**, 1)   (**2**, 3)
Ring Toss   Face Painting

2 − **1** = 1

Since 2 is to the right of 1 on the *x*-axis, the move is 1 unit to the right.

To describe the vertical move, look at the change in the *y*-coordinates.

(1, **1**)   (2, **3**)
Ring Toss   Face Painting

3 − **1** = 2

Since 3 is above 1 on the *y*–axis, the move is 2 units up.

You can describe the move from point (1, 1) to point (2, 3) as 1 unit right and 2 units up.

▶ Reflect

1 Pick two other points on the coordinate plane on the previous page. Use the coordinates of the points to describe a path from one point to the other.

_____
_____
_____
_____
_____

Lesson 30 Graph Points in the Coordinate Plane

Lesson 30 Modeled and Guided Instruction

Learn About Relationships in the Coordinate Plane

Read the problem below. Then explore different ways to represent and use relationships between quantities to solve problems.

> Gabe has $8. He goes to an arcade where each game costs $1. Use equations, a table, and the coordinate plane to represent the relationship between how many games Gabe plays and how much money he has left. How many games can Gabe play if he wants to have exactly $2 when he leaves the arcade?

▶ **Model It** You can use equations to represent the relationship between the quantities in the problem.

Gabe starts with $8. Each game costs $1.

$8 − ($1 × **number of games**) = **dollars left**

$8 − ($1 × **1**) = **$7** left

$8 − ($1 × **2**) = **$6** left

$8 − ($1 × **3**) = **$5** left

$8 − ($1 × **4**) = **$4** left

$8 − ($1 × **5**) = **$3** left

$8 − ($1 × **6**) = **$2** left

▶ **Model It** You can use a table to model the relationship between the quantities.

The table shows how much money Gabe has left after playing different numbers of games.

Number of Games Played	0	1	2	3	4	5	6
Amount of Money Left ($)	8	7	6	5	4	3	2

Connect It Now you will solve the problem from the previous page by representing the relationship between the quantities using the coordinate plane.

2 What quantity in the problem stays the same? _____

What two quantities change? _____

3 Look at the coordinate plane. What quantity is represented on the *x*-axis?

_____ What quantity is represented on the *y*-axis? _____

4 How much money does Gabe have after he plays 0 games? Complete the ordered pair.

(0 games, $_____)

Plot the point in the coordinate plane.

5 Complete these ordered pairs, where *x* is games played and *y* is dollars left.

(1, _____), (2, _____)
(3, _____), (4, _____)
(5, _____), (6, _____)

Plot each ordered pair in the coordinate plane.

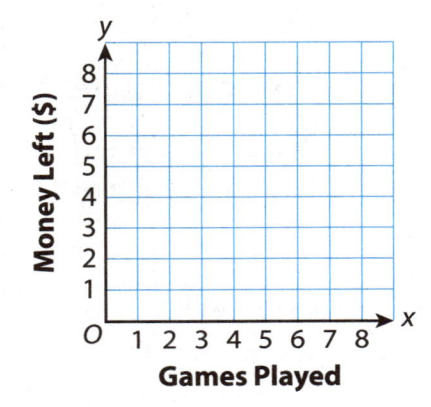

6 How many games can Gabe play and have exactly $2 when he leaves? Which ordered pair shows that? _____

7 How can you use a graph to represent relationships between two sets of numbers?

_____
_____

Try It Use what you just learned about representing relationships between two quantities using the coordinate plane to solve this problem. Show your work on a separate sheet of paper.

8 Suppose that the games at the arcade cost $2 each instead of $1. Write ordered pairs to represent the amounts of money left after playing different numbers of games. Use an "X" to plot each point on the coordinate plane above. How many of these games can Gabe play and have exactly $2 left when he leaves the arcade?

_____

©Curriculum Associates, LLC   Copying is not permitted.

Lesson 30 Graph Points in the Coordinate Plane   **319**

Lesson 30　Modeled and Guided Instruction

Learn About　Solving Problems on a Coordinate Plane

Read the problem below. Then explore different ways of using the coordinate plane to solve a problem.

> Mrs. Sautter writes the following ordered pairs on the board: *M*(2, 1), *A*(2, 5), *T*(5, 5), and *H*(5, 1). The point represented by each ordered pair is a vertex of a rectangle. She lists the vertices in order to name the figure "rectangle *MATH*." What is the perimeter of rectangle *MATH*?

▶ **Picture It** You can picture the rectangle on a coordinate plane.

Plot the points for the vertices of the rectangle. Connect the points to draw the sides of the rectangle. Then shade the rectangle.

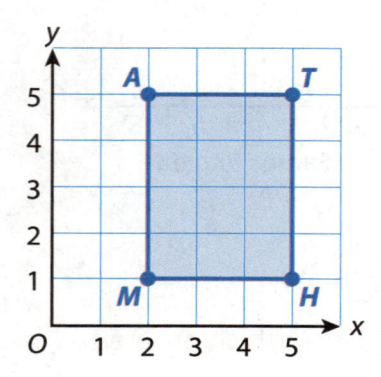

To find the perimeter, count the number of units on each side to find the side lengths.

▶ **Model It** You can find the perimeter of a rectangle with a formula.

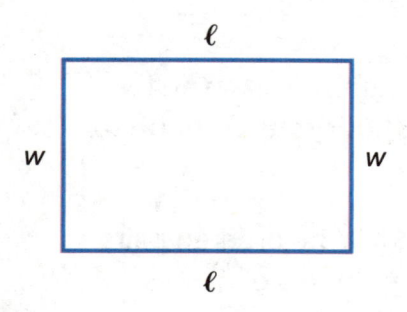

$P = 2\ell + 2w$ or $P = 2 \times (\ell + w)$

320　Lesson 30 Graph Points in the Coordinate Plane　　©Curriculum Associates, LLC　Copying is not permitted.

Connect It Now you will solve the problem from the previous page using your understanding of distance between two points on a number line.

9 What are the ordered pairs for points A and T? _____

Which are different, the x-coordinates or the y-coordinates? _____

Write an expression to find the distance between the points A and T. _____

What is the length of $\overline{AT}$? _____ units

10 Consider how you found the length of $\overline{AT}$ above.

What is the length of $\overline{MA}$? _____ units

Explain how you found this length. _____

_____

_____

11 What is the perimeter of the rectangle? _____ units

Explain how you found the perimeter. _____

_____

12 How can you find the vertical distance between two points on a coordinate plane?

_____

_____

13 How can you find the horizontal distance between two points on a coordinate plane?

_____

_____

Try It Use what you just learned about finding the distance between two points on a number line to solve these problems.

14 What is the vertical distance between (3, 8) and (3, 4)? _____

15 What is the horizontal distance between (4, 2) and (4, 5)? _____

Lesson 30 Graph Points in the Coordinate Plane **321**

Lesson 30 Guided Practice

Practice: Solving Problems on a Coordinate Plane

Study the example below. Then solve problems 16–18.

Example

The graph shows how Jaina's parents determine her weekly allowance. What is the meaning of (3, 5)?

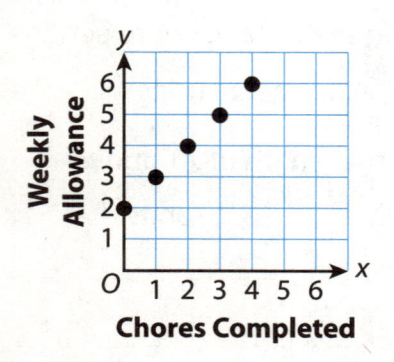

Look at how you can use the graph to solve this problem.

The *x*-axis is labeled "Chores Completed." The *x*-coordinate in (3, 5) is 3. The *y*-axis is labeled "Weekly Allowance." The *y*-coordinate in (3, 5) is 5.

Solution If Jaina completes 3 chores, she will earn $5.

> The labels on the axes tell me what the numbers mean.

> **Pair/Share**
> Explain how to use the graph to find how many chores Jaina must complete to earn $4.

---

16 Plot the points (4, 4), (8, 4), (3, 1), and (7, 1) on the coordinate plane. Use the points to draw two parallel, horizontal segments. Label the endpoints of one segment *A* and *B*. Label the endpoints of the other segment *C* and *D*. What is the distance between points *A* and *B*? What is the distance between points *C* and *D*?

> Do parallel lines ever intersect?

> **Pair/Share**
> What shape could have the 4 points as vertices? Draw the shape on the coordinate plane.

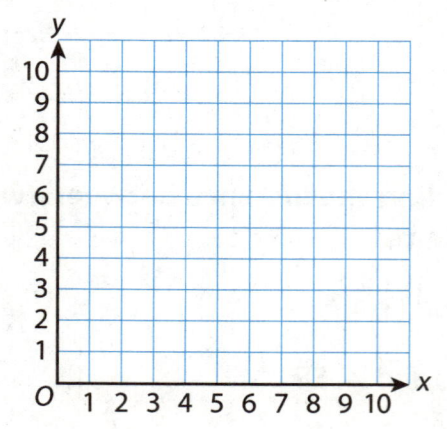

Solution _____

322 Lesson 30 Graph Points in the Coordinate Plane

17 What is the area of rectangle *EFGH* shown on the coordinate plane?

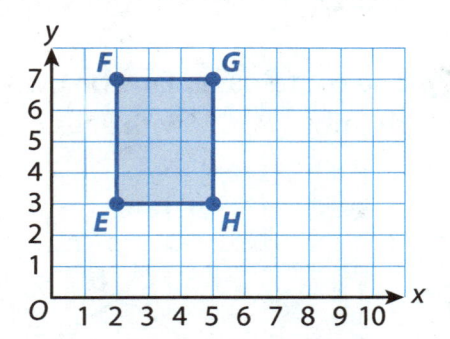

The formula for the area of a rectangle is $A = \ell \times w$.

Show your work.

Pair/Share
Suppose the x-coordinates for points *G* and *H* each increase by 1. What kind of shape would this be? What is the area of the shape?

Solution _____

18 Mr. Palmer uses the coordinate plane to design his bulletin board.

He moves *Rules* 2 units right and 3 units down. What ordered pair represents the new location of *Rules*? Circle the letter of the correct answer.

A (5, 3)

B (1, 9)

C (1, 3)

D (5, 9)

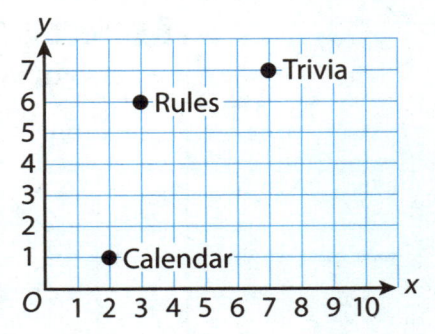

Maya chose **C** as the correct answer. How did she get that answer?

What is the ordered pair for the starting point?

Pair/Share
Does Maya's answer make sense?

©Curriculum Associates, LLC   Copying is not permitted.

Lesson 30 Graph Points in the Coordinate Plane   **323**

Lesson 30 — Independent Practice

Practice: Solving Problems on a Coordinate Plane

Solve the problems.

1 Look at triangle *ABC*.

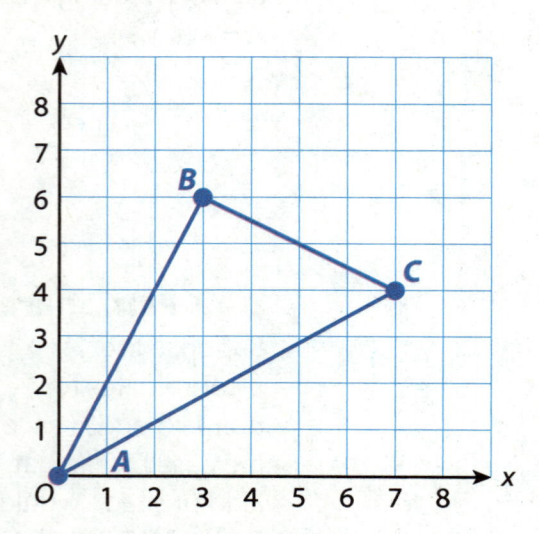

What are the coordinates of points *A*, *B*, and *C*?

A A(1, 1), B(6, 3), C(4, 7)

B A(1, 1), B(3, 6), C(7, 4)

C A(0, 0), B(6, 3), C(4, 7)

D A(0, 0), B(3, 6), C(7, 4)

2 The coordinate plane below shows rhombus *RHOM* with lines of symmetry.

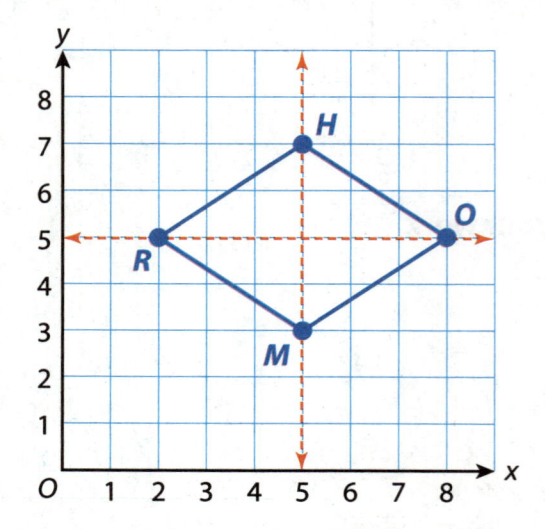

What is the horizontal distance between points *R* and *O*?

_____ units

3 Plot the following three points on the coordinate plane below.

(2, 6)   (7, 4)   (4, 0)

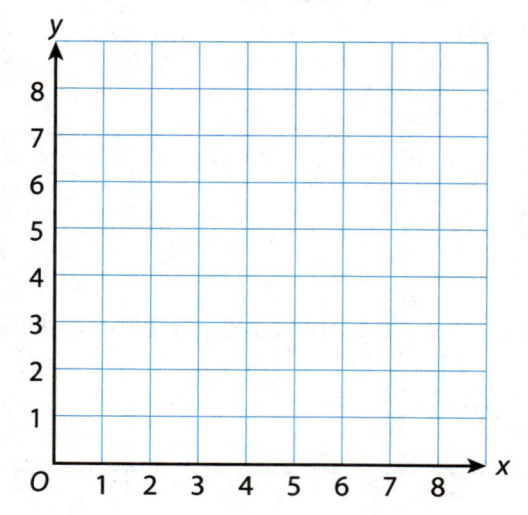

324 Lesson 30 Graph Points in the Coordinate Plane

4 The coordinate plane shows the relationship between the amount of water in the bathtub and the amount of time the faucet has been turned on.

How many minutes must the faucet be turned on for the bathtub to hold 8 gallons of water?

A 16 minutes

B 8 minutes

C 4 minutes

D 2 minutes

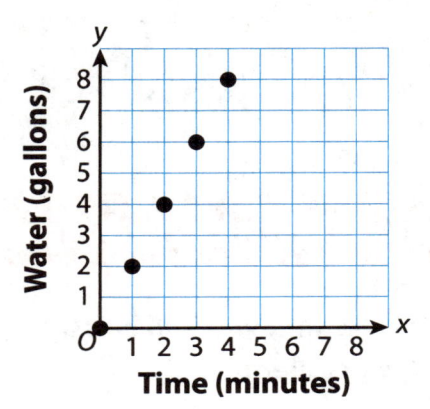

5 Part A The area of rectangle *RSTU* is 20 square units. Draw rectangle *RSTU* on the coordinate plane below with one vertex at the origin.

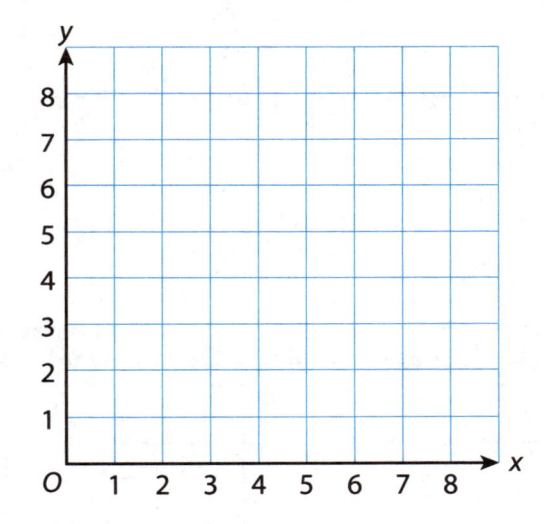

Part B Write the ordered pairs for the points *R*, *S*, *T*, and *U*. _____

Part C How do you know that the area of rectangle *RSTU* is 20 square units?

✓ **Self Check** Go back and see what you can check off on the Self Check on page 309.

Lesson 30 Graph Points in the Coordinate Plane **325**

Lesson 31 — Introduction
Classify Quadrilaterals

NC.5.G.3

Use What You Know

In this lesson, you will classify quadrilaterals based on their properties. Take a look at this problem.

> Arrange the quadrilaterals below so that a quadrilateral can also be called by the name of the quadrilateral before it. Order them from left to right.
>
> A   B   C

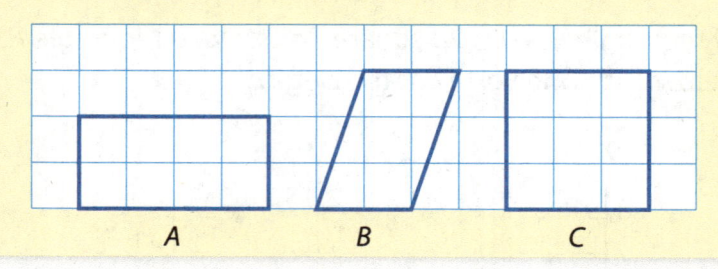

a. Complete the table below. Put a check in each box if the quadrilateral has the property listed.

Property	Quadrilateral A	Quadrilateral B	Quadrilateral C
4 sides			
2 pairs of parallel sides			
2 pairs of sides of equal length			
4 right angles			
4 sides of equal length			

b. Write the most specific name for each polygon from the list below.

quadrilateral   parallelogram   rectangle   square

A: _____   B: _____   C: _____

c. How would you arrange the polygons so each shape has all the properties of the shape(s) before it? _____

326 Lesson 31 Classify Quadrilaterals

▶▶ Find Out More

Shapes can be classified according to their properties. When you order categories of quadrilaterals by their properties, you put them in a **hierarchy**. A hierarchy organizes categories from the most to least general. One model you can use to show a hierarchy is a Venn diagram.

A Venn diagram can show categories and subcategories. This Venn diagram shows that squares have all the properties that rectangles have, plus more. This means all squares are also rectangles. A square is also a parallelogram and a quadrilateral.

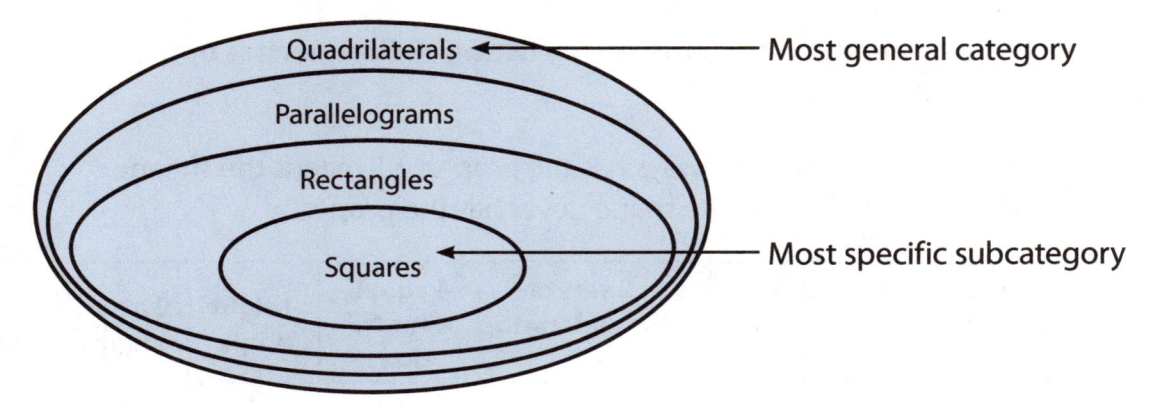

Quadrilaterals ← Most general category
Parallelograms
Rectangles
Squares ← Most specific subcategory

You can also use a flow chart to show the hierarchy of quadrilaterals. The most general category is at the left, while the most specific is at the right. This means that a figure that belongs in one category also belongs in all categories to the left.

Quadrilaterals → Parallelograms → Rectangles → Squares

▶ Reflect

1 How are the flow chart and the Venn diagram alike? How are they different?

_____

_____

_____

_____

_____

Lesson 31 Classify Quadrilaterals

Lesson 31 **Modeled and Guided Instruction**

Learn About: Ordering Shapes in a Hierarchy

Read the problem below. Then explore different ways to classify figures in a hierarchy.

> Use a tree diagram to classify the following quadrilaterals from most general to most specific: parallelogram, rectangle, square, rhombus, trapezoid. Arrange each branch from most general to most specific.

▶ **Model It** You can understand the problem by listing the properties of the quadrilaterals' sides and angles.

In the chart below, *A* means the shape always has the property. *S* means the shape sometimes has the property. *N* means the shape never has the property.

	At least 1 set of parallel sides	2 sets of parallel sides	2 sets of equal-length sides	4 equal-length sides	1 right angle	2 acute angles	4 right angles
Parallelogram	A	A	S	S	S	S	S
Trapezoid	A	S	S	S	S	S	S
Rectangle	A	A	A	S	A	N	A
Rhombus	A	A	A	A	S	S	S
Square	A	A	A	A	A	N	A

▶ **Model It** You can represent the problem with a tree diagram.

A tree diagram can also be used to show a hierarchy. Put the most general category as the top branch. Then put the more specific subcategories as the branches.

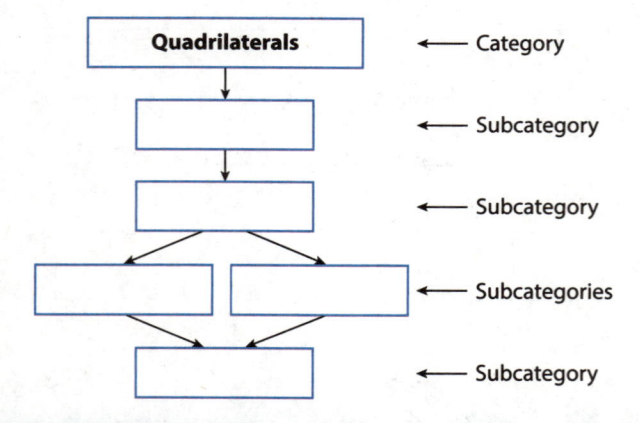

Quadrilaterals ← Category
　↓
　[ ] ← Subcategory
　↓
　[ ] ← Subcategory
　↙ ↘
[ ]　[ ] ← Subcategories
　↘ ↙
　[ ] ← Subcategory

328 Lesson 31 Classify Quadrilaterals

▶ **Connect It** Now you will solve the problem from the previous page by using the chart to complete a tree diagram.

2 Why is "Quadrilaterals" in the top row of the tree diagram? _____
_____

3 Look back at the chart from the first *Model It*. Is a shape with more *A* entries more specific or more general than a shape with more *S* entries? Explain.
_____
_____
_____

4 Which category has the fewest *A* entries on the chart? _____
Which has the most? _____
Fill in those categories on the tree diagram.

5 Fill in the tree diagram with the remaining categories.

6 How can you use a tree diagram to order figures? _____
_____
_____
_____

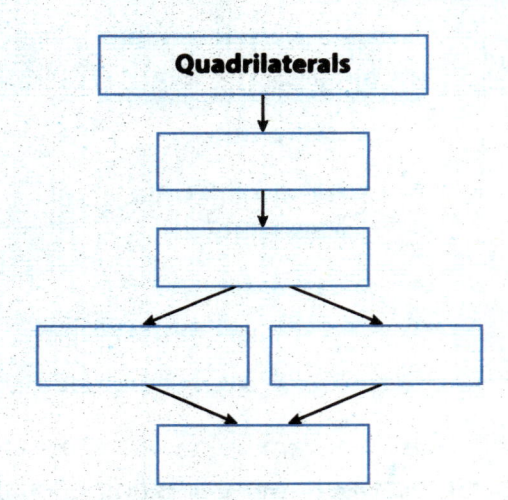

▶ **Try It** Use what you learned about ordering figures in a hierarchy to solve this problem.

7 Complete the Venn diagram at the right to show the hierarchy of quadrilaterals, rectangles, trapezoids, and squares.

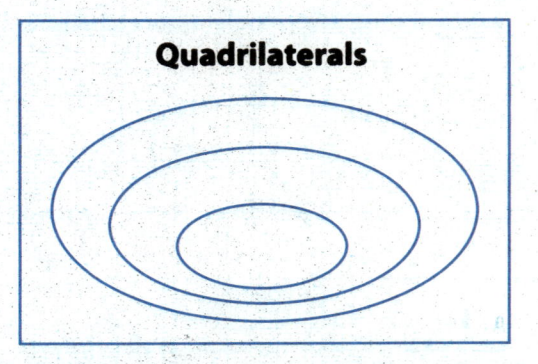

Lesson 31 Classify Quadrilaterals **329**

Lesson 31 Guided Practice

Practice: Classifying Quadrilaterals

Study the example below. Then solve problems 8–10.

Example

Create a Venn diagram to show the hierarchy of quadrilaterals, rectangles, rhombuses, and squares.

Look at how you could show your work using a Venn diagram.

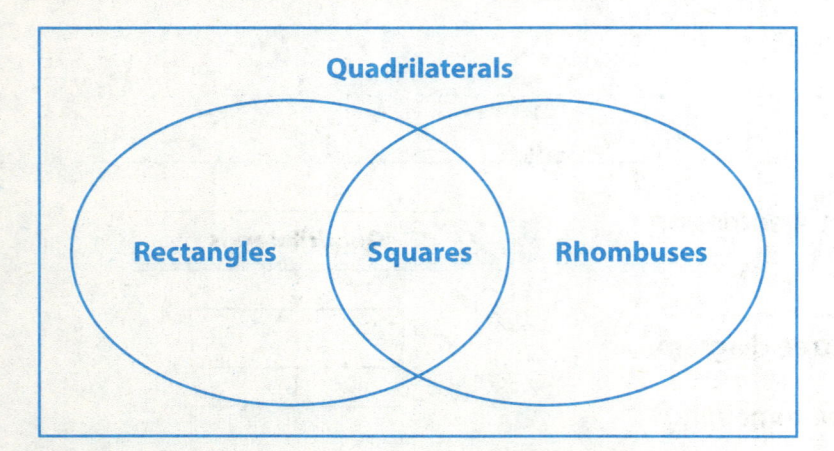

Quadrilaterals
- Rectangles | Squares | Rhombuses

Overlapping circles can show categories that are sometimes the same but not always the same.

Pair/Share
Recreate the hierarchy with a tree diagram.

---

8 Look at the tree diagram below. Write a statement about the relationship between trapezoids and parallelograms.

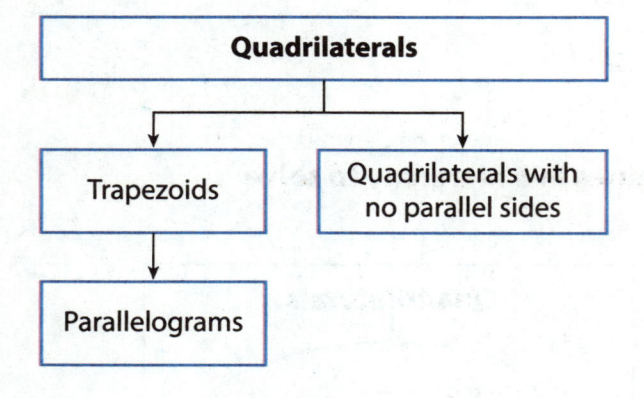

Quadrilaterals
→ Trapezoids → Parallelograms
→ Quadrilaterals with no parallel sides

Which type of quadrilateral is the most specific?

Pair/Share
Write a statement about the relationship between parallelograms and quadrilaterals with no parallel sides.

Solution _____

_____

330 Lesson 31 Classify Quadrilaterals

©Curriculum Associates, LLC Copying is not permitted.

9 Create a Venn diagram to show the hierarchy of the quadrilaterals described in the chart.

Quadrilateral	Description
Trapezoid	quadrilateral with at least 1 pair of parallel sides
Square	parallelogram with 4 sides of equal length
Parallelogram	quadrilateral with 2 pairs of parallel sides

"At least 1" means 1 or more.

Pair/Share
Draw one example of a quadrilateral in each separate category of your Venn diagram.

10 Look at the flow chart below.

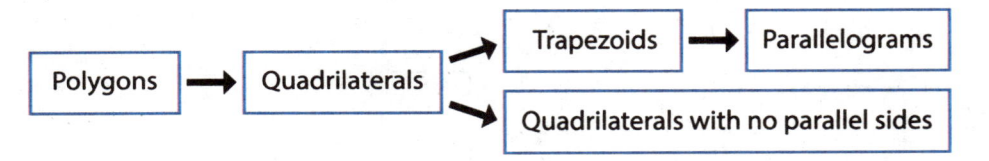

Polygons → Quadrilaterals → Trapezoids → Parallelograms
Quadrilaterals → Quadrilaterals with no parallel sides

Which statement is true? Circle the letter of the correct answer.

A   A polygon is always a quadrilateral.

B   All quadrilaterals are polygons.

C   All quadrilaterals are trapezoids and parallelograms.

D   A parallelogram is not a polygon.

Brad chose **C** as the correct answer. How did he get that answer?

_____

_____

_____

The flow chart is like a tree diagram. But the arrows show that the hierarchy moves from left to right instead of top to bottom.

Pair/Share
Does Brad's answer make sense?

Lesson 31 Classify Quadrilaterals   **331**

Lesson 31 — Independent Practice

Practice: Classifying Quadrilaterals

Solve the problems.

1 Look at the shape below.

Which is a correct classification for this shape from LEAST specific to MOST specific?

A polygon, quadrilateral, rectangle

B quadrilateral, parallelogram, square

C polygon, quadrilateral, square

D quadrilateral, rectangle, square

2 Classify the quadrilaterals shown below as "rectangle," "square," or "rhombus." Sides that are the same length are marked with a slash. Draw the quadrilaterals in the correct column of the table. If a quadrilateral fits more than one classification, draw it in all the columns that apply.

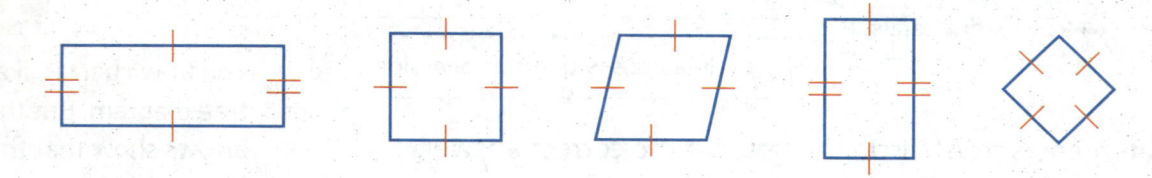

Square	Rectangle	Rhombus

332   Lesson 31 Classify Quadrilaterals

3 Look at the flow chart below.

Quadrilaterals → Trapezoids → Parallelograms → Rectangles → Squares

Part A Draw an example of a trapezoid that is not a parallelogram.

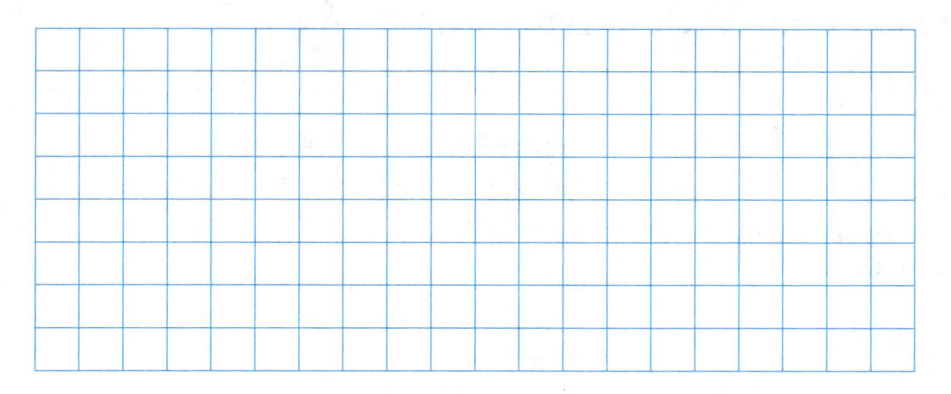

Part B Explain how trapezoids relate to parallelograms.

_____
_____
_____
_____

Part C Can you use the term "parallelogram" to describe a rectangle? Explain your reasoning.

_____
_____
_____
_____

✓ **Self Check** Go back and see what you can check off on the Self Check on page 309.

Lesson 32 · Introduction
Understand Properties of Quadrilaterals

NC.5.G.3

Think It Through

How do we group quadrilaterals into categories?

Quadrilaterals are grouped into categories by their **attributes**, or properties, such as the number of sides or angles, the side lengths, and the angle measures. All quadrilaterals in the same category share certain properties. Some properties of quadrilaterals are described in the table below.

Property	Description	Example
Regular	all sides of equal length and all angles of equal measure	
Irregular	at least 1 side and 1 interior angle are not equal in measure to the other sides and angles	
Right	at least 1 pair of perpendicular sides	
Parallel sides	at least 1 pair of opposite sides that will never intersect, no matter how far they are extended	

Think Can a quadrilateral be categorized in more than one way?

A quadrilateral is defined as a polygon with 4 sides. So any shape with 4 sides can be called both a polygon and a quadrilateral. If the quadrilateral has two pairs of parallel sides, then it can also be called a parallelogram.

Every parallelogram is a quadrilateral because every parallelogram has 4 sides. But not all quadrilaterals are parallelograms because not all quadrilaterals have two pairs of parallel sides.

✏️ **Shade** a polygon above that can be named both a quadrilateral and parallelogram.

Think How can you show the relationships among quadrilaterals with a diagram?

A Venn diagram is a useful tool for organizing categories of quadrilaterals that share properties.

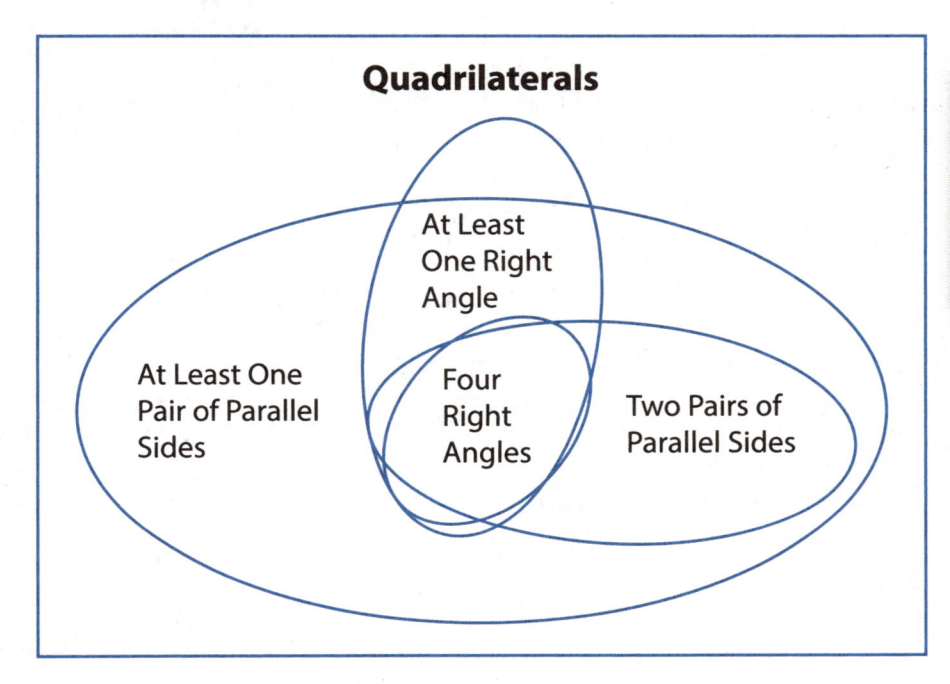

Quadrilaterals

- At Least One Right Angle
- At Least One Pair of Parallel Sides
- Four Right Angles
- Two Pairs of Parallel Sides

The Venn diagram shows that a quadrilateral with four right angles will always have two pairs of parallel sides.

Notice that the category "At Least One Right Angle" partially overlaps several other categories, including "At Least One Pair of Parallel Sides." That means that a quadrilateral with one or more right angles may have zero, one, or two pairs of parallel sides. Notice, also, that the category "Two Pairs of Parallel Sides" is fully enclosed within "At Least One Set of Parallel Sides."

The category "Four Right Angles" has the smallest circle, and it is enclosed fully within the overlap between the categories "At Least One Right Angle," "At Least One Pair of Parallel Sides," and "Two Pairs of Parallel Sides."

Reflect

1 What does it mean that the Venn diagram shows the category "Four Right Angles" completely inside "Two Pairs of Parallel Sides"?

_____

_____

Lesson 32 Guided Instruction

Think About: Properties Shared by Quadrilaterals

Let's Explore the Idea A Venn diagram can help you understand the properties shared by categories of quadrilaterals.

2 The Venn diagram shows categories of quadrilaterals with different properties. Write the name of each category that fits the description.

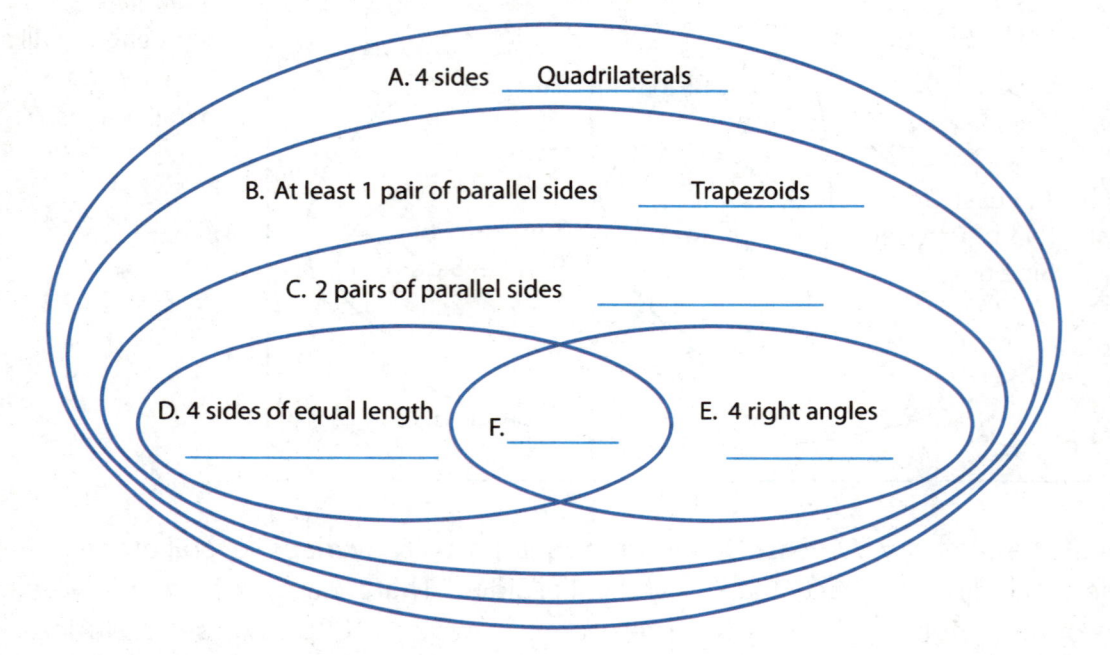

- A. 4 sides — Quadrilaterals
- B. At least 1 pair of parallel sides — Trapezoids
- C. 2 pairs of parallel sides — _____
- D. 4 sides of equal length — _____
- E. 4 right angles — _____
- F. _____

3 Use the Venn diagram to fill in the table below.

Category	Properties	Name
A	4 sides	Quadrilaterals
B	4 sides, at least 1 pair of parallel sides	Trapezoids
C	4 sides, 2 pairs of parallel sides	
D	4 sides, 2 pairs of parallel sides, 4 sides of equal length	
E		
F		

Lesson 32 *Understand* Properties of Quadrilaterals

Let's Talk About It Use the Venn diagram to help you understand how properties are shared by categories of quadrilaterals.

4 Is every property of parallelograms also a property of all rectangles? _____

Is every property of rectangles also a property of all parallelograms? _____

Explain what the Venn diagram shows about the relationship between rectangles and parallelograms. _____

Classify each inference statement as *true* or *false*. If false, explain.

5 The opposite angles of any parallelogram have the same measure. Therefore, the opposite angles of any rhombus have the same measure. _____

6 The diagonals of any square are the same length. Therefore, the diagonals of any rhombus are the same length. _____

Try It Another Way The flow chart below shows another way to think about how quadrilaterals are categorized.

Quadrilaterals → Trapezoids → Parallelograms → Rectangles → Squares
                                              → Rhombuses →

Use the flow chart to describe the statements as *true* or *false*.

7 In every rectangle the two diagonals have the same length. Therefore, in every parallelogram the two diagonals must have the same length. _____

8 Every rhombus has at least 2 lines of symmetry. Therefore, every square has at least 2 lines of symmetry. _____

Lesson 32 *Understand* Properties of Quadrilaterals **337**

Lesson 32 Guided Practice
Connect: Ideas About Properties of Quadrilaterals

Talk through these problems as a class. Then write your answers below.

9 Categorize All quadrilaterals are either **convex** or **concave**. A convex quadrilateral has all interior angles less than 180°. A rhombus is an example of a convex quadrilateral. A concave quadrilateral has at least 1 interior angle greater than 180°. The quadrilateral below is an example of a concave quadrilateral.

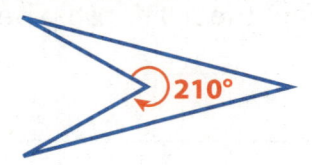

210°

Categorize concave quadrilaterals, convex quadrilaterals, trapezoids, and rectangles in a Venn diagram. Draw an example of each quadrilateral in the diagram.

10 Explain Nadriette said that a rectangle can never be called a trapezoid. Explain why Nadriette's statement is incorrect. _____

11 Create Describe the properties of a shape that is both a rectangle and a rhombus. Name the shape and use the grid below to draw an example.

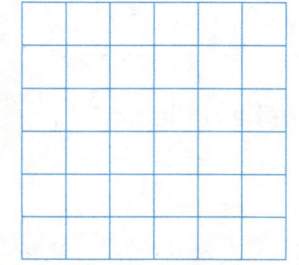

338 Lesson 32 *Understand* Properties of Quadrilaterals

Lesson 32 — Independent Practice

Apply: Ideas About Properties of Quadrilaterals

12 Put It Together Use what you have learned about classifying quadrilaterals to complete this task.

Part A Create a tree diagram to show the following types of quadrilaterals: trapezoid, parallelogram, square, rhombus, rectangle. Make sure to include the category "Quadrilateral." Use information in the table to help you.

Quadrilateral	Types of Angles	Types of Sides
Trapezoid	can include any types of angles	At least one set of parallel sides
Parallelogram	2 sets of congruent angles	2 sets of congruent parallel sides
Square	4 right angles	4 congruent sides, 2 sets of parallel sides
Rhombus	2 sets of congruent angles	4 congruent sides, 2 sets of parallel sides
Rectangle	4 right angles	2 sets of congruent parallel sides

Part B Write a statement that is always true about the relationship between squares and rhombuses.

Part C Write a statement that is sometimes true about the relationship between squares and rectangles.

Unit 5 MATH IN ACTION

Introduction
Work with Geometry and Coordinates

SMP1 Make sense of problems and persevere in solving them.

Study an Example Problem and Solution

In this lesson, you will use what you know about shapes and coordinate grids to solve real-world problems. Look at this problem and one solution.

Octagon Trap

Max is working on a new video game, Shape Shake-Up. He uses a coordinate grid to represent the screen. It helps him decide where to place graphics. Read one of Max's ideas.

Game Idea
- A shape that looks like a C traps players.
- The shape's perimeter is 14 to 16 units. Its area is 6 to 8 square units.
- The shape is located more than 2 units above the x-axis and more than 2 units to the right of the y-axis.

Draw a shape on a coordinate grid that works with Max's game idea and explain why it works. Label each vertex with a coordinate pair.

Read the sample solution on the next page. Then look at the checklist below. Find and mark parts of the answer that match the checklist.

✏️ Problem-Solving Checklist

- ☐ Tell what is known.
- ☐ Tell what the problem is asking.
- ☐ Show all your work.
- ☐ Show that the solution works.

a. **Circle** something that is known.
b. **Underline** something that you need to find.
c. **Draw a box around** what you do to solve the problem.
d. **Put a checkmark** next to the part that shows the solution works.

340 Unit 5 Math in Action Work with Geometry and Coordinates ©Curriculum Associates, LLC Copying is not permitted.

Max's Solution

▷ **I know** the choices for area and perimeter of the shape. I need to find the lengths of sides that can make an area and perimeter that work.

▷ **I can try** different lengths for the sides.
There are 8 sides.
Their lengths must have a sum of 14 to 16.
It looks like I should try numbers close to 2.

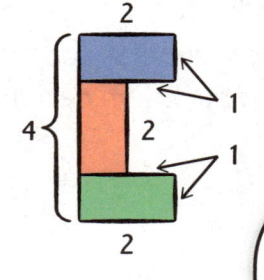

▷ **I'll check** that the perimeter works.
4 + (3 × 2) + (1 × 4) = 14

▷ **I'll check** that the area works.
(2 × 1) + (2 × 1) + (2 × 1) = 6

The shape has a perimeter of 14 units and an area of 6 square units.

▷ **Now I can plot the shape on a coordinate grid.**
I'll put the bottom left vertex at (3, 3) so I know it's more than 2 units from each axis.

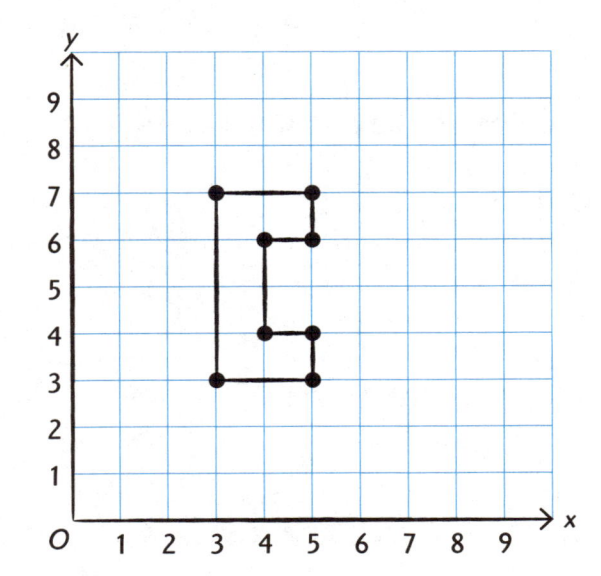

Hi, I'm Max. Here's how I solved this problem.

First I tried a shape that was 5 units wide and 5 units tall, but that was much too big.

After I put a corner at (3,3) I counted right 2, up 1, left 1, up 2, and so on to draw the whole shape.

Unit 5 Math in Action   Work with Geometry and Coordinates

Unit 5 Math in Action — Modeled and Guided Instruction

Try ▶ Another Approach

There are many ways to solve problems. Think about how you might solve the Octagon Trap problem in a different way.

Octagon Trap

Max is working on a new video game, Shape Shake-Up. He uses a coordinate grid to represent the screen. It helps him decide where to place graphics. Read one of Max's ideas.

Game Idea
- A shape that looks like a C traps players.
- The shape's perimeter is 14 to 16 units. Its area is 6 to 8 square units.
- The shape is located more than 2 units above the x-axis and more than 2 units to the right of the y-axis.

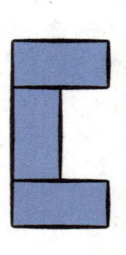

Draw a shape on a coordinate grid that works with Max's game idea and explain why it works. Label each vertex with a coordinate pair.

▶ **Plan It** Answer these questions to help you start thinking about a plan.

A. How can you make a different "C" shape?

_____
_____

B. What are some different locations you can use on the grid?

_____
_____

▶ **Solve It** Find a different solution for the Octagon Trap problem. Show all your work on a separate sheet of paper.

You may want to use the problem-solving tips to get started.

Problem-Solving Tips

- **Tools** You might want to use . . .
  - grid paper.
  - dot paper.

- **Word Bank**

perimeter	total	axis
area	units	coordinate pair
sum	square units	

- **Sentence Starters**
  - To find the perimeter _____
  - The area of my shape _____

Problem-Solving Checklist

Make sure that you . . .
☐ tell what you know.
☐ tell what you need to do.
☐ show all your work.
☐ show that the solution works.

▶ **Reflect**

Use Mathematical Practices As you work through the problem, discuss these questions with a partner.

- **Use Models** How can a sketch help you solve this problem?
- **Use Tools** Will a ruler be helpful for this problem? Explain how it can help, or explain why not.

Unit 5 Math in Action — Guided Practice

Discuss: Models and Strategies

Read the problem. Write a solution on a separate sheet of paper. Remember, there can be lots of ways to solve a problem!

Rectangle Maze

One screen in Max's game has a maze of rectangles. Players try to move through the maze. The rectangles start in a certain location, but they move both left and right and up and down. Max will decide the starting location for each rectangle.

Max's Notes
- Show 5 rectangles.
- Put one rectangle near the middle of the screen.
- Put other rectangles on all 4 sides of the middle rectangle.
- Other rectangles must have an area that is $\frac{1}{2}$ of the area or perimeter that is $\frac{1}{2}$ of the perimeter of the middle rectangle.
- Each shape must be a distance of at least 1 unit from the middle rectangle on all sides.

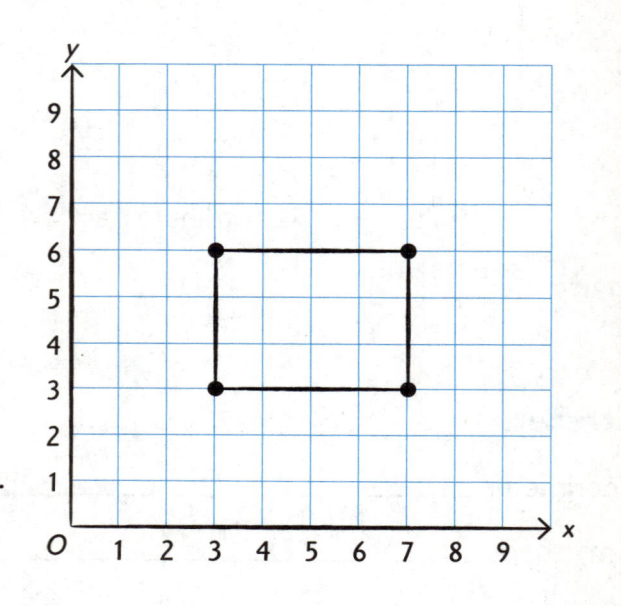

Where should Max put the other 4 rectangles?

▶ **Plan It and Solve It** Find a solution to the Rectangle Maze problem.

Use the Rectangle Maze Activity Sheet and the information in Max's notes.
- Draw the other 4 rectangles.
- Label the vertices of all shapes with coordinate pairs.
- Explain why your shapes work with Max's notes.

Problem-Solving Tips

- **Questions**
  - What areas can your rectangles have?
  - What perimeters can your rectangles have?

- **Word Bank**

area	distance	coordinate pair
perimeter	unit	coordinate grid

Problem-Solving Checklist

Make sure that you . . .
- ☐ tell what you know.
- ☐ tell what you need to do.
- ☐ show all your work.
- ☐ show that the solution works.

▶ **Reflect**

Use Mathematical Practices As you work through the problem, discuss these questions with a partner.

- **Be Precise** How can you find the area and perimeter of a rectangle on a coordinate grid?
- **Use Models** What expressions can you write to find half of the perimeter and half of the area of the given rectangle?

Unit 5 Math in Action   **Independent Practice**

Persevere ▶ On Your Own

Read the problems. Write a solution on a separate sheet of paper. Remember, there are many different ways to solve a problem.

The Plunging Parallelogram

In Max's game, a parallelogram starts at the top of the screen and zigzags down to the bottom. Read Max's design notes.

Max's Design Notes
- The parallelogram has all right angles.
- The sides are at least 2 units long.
- At the start, one vertex is located at (10, 10).
- At the end, one vertex is located at (0, 0).

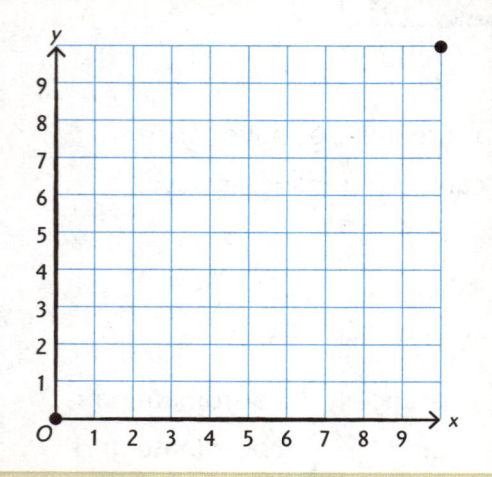

What points should Max use as the parallelogram's vertices?
What path can the shape take to get to the bottom of the screen?

▶ Solve It Help Max draw the parallelogram.
- Plot points and draw the parallelogram at its starting point.
- Give the coordinates of each vertex.
- Plot points and draw the parallelogram at its ending point.
- Describe a path the shape could follow to get to the bottom of the screen. Use directional words and numbers of units in your description.

▶ Reflect

Use Mathematical Practices After you complete the task, choose one of these questions to discuss with a partner.
- **Critique Reasoning** Does the path your partner described work? Why or why not?
- **Be Precise** What are all the different names that can be used to describe Max's shape?

Hierarchy Hit

At three times in Max's game a quadrilateral appears on the screen. Players smash the quadrilateral to score bonus points. When a shape is hit, it morphs into a different quadrilateral. It continues to morph into other quadrilaterals according to the rules below.

Hierarchy Hit Rules
- The morphing can go from the more general to more specific shape. Or it can go from the more specific to more general shape.
- There are at least three different quadrilaterals in the morphing sequence.

▶ **Solve It** Help Max design the hierarchy.
- Find quadrilaterals that will fit with the rules.
- Make a flow chart that shows a hierarchy of quadrilaterals and their names.
- Draw an example of each quadrilateral.
- Explain why the order of shapes shows a hierarchy.

▶ **Reflect**

Use Mathematical Practices After you complete the task, choose one of these questions to discuss with a partner.

- **Use Models** How does the flow chart you made show the hierarchy of shapes? Explain.
- **Be Precise** How does the most specific shape you drew fit in all of the other categories? Explain.

Unit 5 Assessment
Interim Assessment

Solve the problems.

1. Jack, Chris, Ryan, Peter, and Sam all attend Laurelleaf Elementary School. The coordinate plane below shows the locations of their houses and the school.

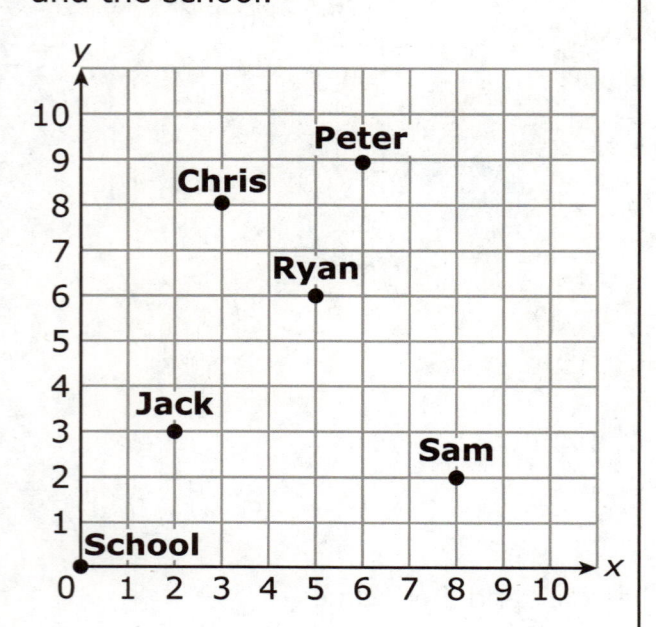

   What x-coordinate describes the location of Ryan's house?

   A   2
   B   3
   C   5
   D   6

2. Which figure shown below is a rectangle?

   A 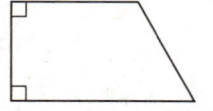
   B 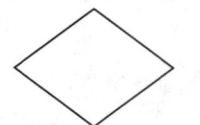
   C 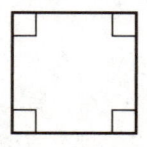
   D

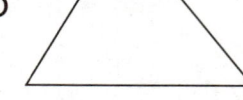

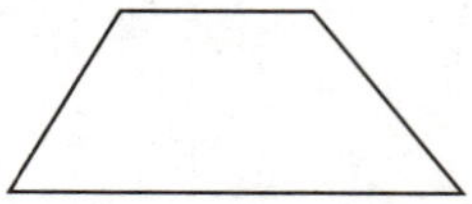

3. Look at rectangle ABCD on the coordinate plane below.

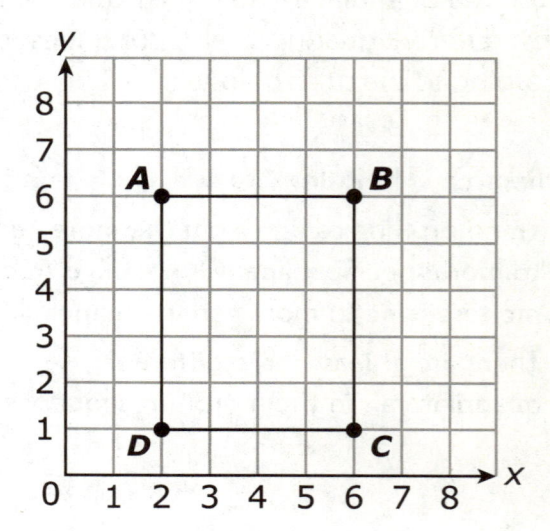

   Which statement is true?

   A   The coordinates of point A are (6, 2).

   B   The coordinates of point D are (2, 1).

   C   The area of rectangle ABCD is 36 square units.

   D   The perimeter of rectangle ABCD is 18 units.

4   Which statement is true based on the diagram at the right?

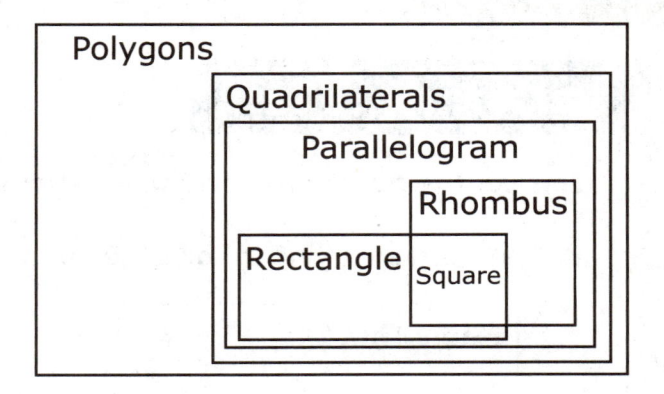

   A   All squares are rhombuses.

   B   All rectangles are squares.

   C   All quadrilaterals are parallelograms.

   D   All polygons are quadrilaterals.

5   Which quadrilateral is a rectangle?

   A

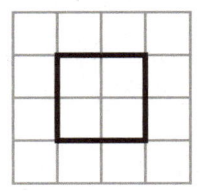

   B

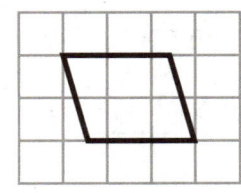

   C

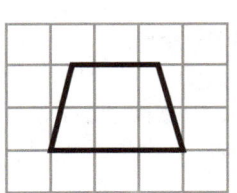

   D

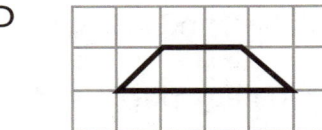

6   The table shows properties of parallelograms and rhombuses.

All parallelograms have:	All rhombuses have:
4 sides	4 sides of equal length
4 vertices	4 vertices
2 pairs of parallel sides	2 pairs of parallel sides

   Use the information in the table. Which statement below is true?

   A   All parallelograms are rhombuses.

   B   All rhombuses are parallelograms.

   C   Parallelograms are never rhombuses.

   D   Rhombuses are never parallelograms.

Unit 5 Interim Assessment   349

Unit 5 Interim Assessment continued

Performance Task

Answer the questions and show all your work on separate paper.

Ben and Cari are trying to solve the puzzle below.

> **Name That Shape**
>
> 1. The figure is a quadrilateral.
>
> 2. The figure has a pair of parallel sides.
>
> 3. Three of the vertices are located at (1, 2), (3, 5), and (7, 5) on a coordinate plane.
>
> What type of figure is it? Where is the fourth vertex?

Checklist
Did you . . .
☐ make a list?
☐ draw the figure on a coordinate grid?
☐ check your vocabulary?

Ben thinks there's only one solution to the puzzle. Cari says there's more than one possible answer. Who do you think is right? Explain your reasoning. Then solve the puzzle. If you agree with Cari, find at least two possibilities for the fourth vertex.

Now create your own geometry puzzle with 3 clues. Make sure your puzzle has only one answer and that it cannot be answered with only 2 clues.

▶ Reflect

Use Mathematical Practices After you complete the task, choose one of the following questions to answer.

- **Use Structure** How did you use the clues to solve the puzzle?

- **Be Precise** What did you think about when writing your own puzzle?

Glossary

A

AM morning, or the time from midnight until noon.

acute angle an angle with a measure less than 90°.

acute triangle a triangle that has three acute angles.

addend a number being added.

angle two rays that share an endpoint.

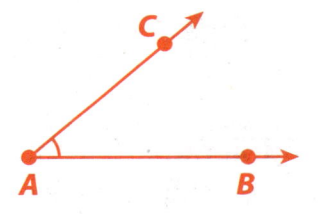

area the amount of space inside a closed, two-dimensional figure. Area is measured in square units such as square centimeters.

array a set of objects grouped in equal rows and equal columns.

associative property of addition Changing the grouping of three or more addends does not change the sum.

(2 + 3) + 4 = 2 + (3 + 4)

associative property of multiplication Changing the grouping of three or more factors does not change the product.

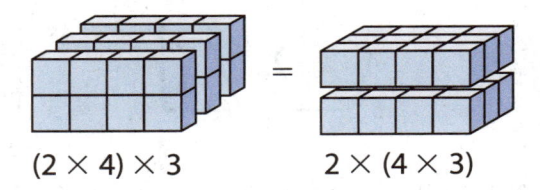

(2 × 4) × 3 = 2 × (4 × 3)

attribute a characteristic of an object. Attributes of a shape include the number of sides and the length of the sides.

B

base ten the number system we use everyday using ten digits. The value of a digit depends on its place value. The value of each place is 10 times the value of the place to the right.

benchmark fraction a common fraction that you might compare other fractions to. For example, $\frac{1}{2}$ can be used as a benchmark.

C

capacity the amount of liquid a container can hold. Capacity is measured in the same units as liquid volume.

centimeter (cm) a unit of length in the metric system. Your little finger is about 1 centimeter across. 100 centimeters is equivalent to 1 meter.

Glossary

closed figure a two-dimensional figure that begins and ends at the same point.

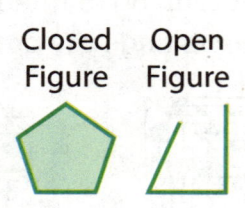

Closed Figure Open Figure

column a top-to-bottom line of objects in an array.

common denominator a number that is a common multiple of the denominators of two or more fractions.

commutative property of addition Changing the order of the addends does not change the sum.

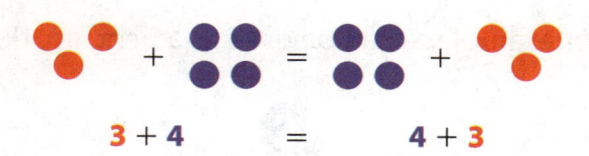

3 + 4 = 4 + 3

commutative property of multiplication Changing the order of the factors does not change the product.

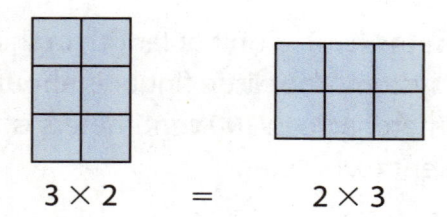

3 × 2 = 2 × 3

compare to decide if one number is greater than (>), less than (<), or equal to (=) another number. Also, to tell how much more one number is than another number, or how many times as many one number is than another number.

compose to make by combining parts.

composite number a number greater than 0 that has more than one pair of factors.

concave quadrilateral a quadrilateral with at least one inside angle with a measure greater than 180°.

convert to change from one measurement unit to another.

convex quadrilateral a quadrilateral with all inside angles with measures less than 180°.

coordinate plane a two-dimensional space formed by two perpendicular number lines called axes.

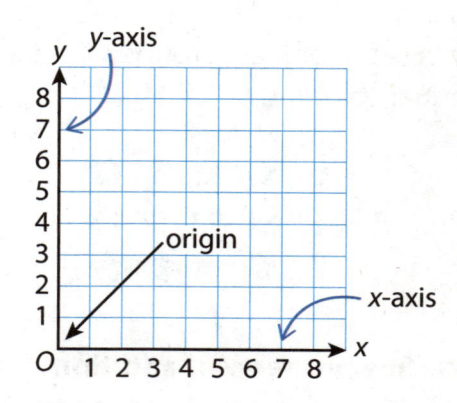

corresponding terms (in two related patterns) the numbers that have the same position. For example, the second term in one pattern and the second term in a related pattern are corresponding terms.

352 Glossary

Glossary

cubic unit a cube with edge lengths of 1 unit. Cubic units are used to measure the volume of a solid figure.

cup (c) a unit of liquid volume in the customary system. Four cups is equivalent to 1 quart.

customary system the measurement system commonly used in the United States. It measures length in inches, feet, yards, and miles; liquid volume in cups, pints, quarts, and gallons; and weight in ounces and pounds.

D

data information, often numerical information such as a list of measurements.

decimal a number containing a decimal point that separates the ones place from the tenths place.

decimal point the period or dot used in a decimal that separates the ones place from the tenths place.

decompose to break up into parts.

degree (°) a unit used to measure angles. There are 360° in one circle.

denominator the number below the line in a fraction. It tells how many equal parts are in the whole.

$$\frac{2}{5} \leftarrow \text{denominator}$$

difference the result of subtraction.

digit a symbol used to write numbers. The digits are 0, 1, 2, 3, 4, 5, 6, 7, 8, and 9.

dimension length in one direction. A figure may have one, two, or three dimensions.

distribution (of data) how spread out or clustered pieces of data are.

distributive property When one of the factors of a product is written as a sum, multiplying each addend by the other factor before adding does not change the product.

$$2 \times (3 + 6) = (2 \times 3) + (2 \times 6)$$

divide to separate into equal groups.

dividend the number that is divided in a division problem.

division an operation used to separate a number of items into equal-sized groups.

divisor the number you divide by in a division problem.

E

elapsed time the time that has passed between a start time and an end time.

equal (=) having the same value, same size, or same amount.

equation a mathematical statement that uses an equal sign (=) to show that two expressions have the same value.

Glossary 353

Glossary

equilateral triangle a triangle that has all three sides the same length.

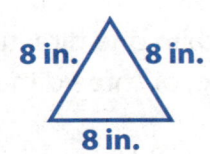

8 in. 8 in.
8 in.

equivalent fractions two or more fractions that have the same value. They name the same part of a whole and the same point on a number line.

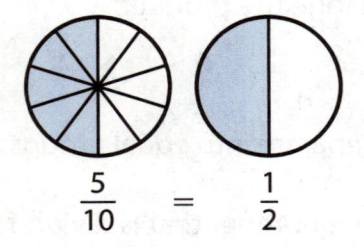

$\frac{5}{10} = \frac{1}{2}$

estimate (noun) a close guess made using math thinking.

estimate (verb) to make a close guess based on math thinking.

evaluate to find the value of an expression.

even number a whole number that ends in the digit 0, 2, 4, 6, or 8. Even numbers have 2 as a factor.

expanded form the way a number is written to show the place value of each digit.

$254.3 = 200 + 50 + 4 + \frac{3}{10}$

expression numbers or unknowns combined with operation symbols. For example, $5 + a$ or 3×6.

F

fact family a group of related math facts that all use the same numbers. The group of facts shows the relationship between addition and subtraction, or between multiplication and division.

$5 \times 4 = 20$
$4 \times 5 = 20$
$20 \div 4 = 5$
$20 \div 5 = 4$

factor a number that is multiplied.

factor pair two numbers that are multiplied together to give a particular product. For example, 1 and 12, 2 and 6, and 3 and 4 are all factor pairs of 12.

factors of a number whole numbers that multiply together to get the given number.

foot (ft) a unit of length in the customary system. One foot is equivalent to 12 inches.

Glossary

formula a mathematical relationship that is expressed in the form of an equation. For example, the formula for the area of a rectangle is $A = \ell \times w$.

fraction a number that names equal parts of a whole; a fraction names a point on the number line and can also represent the division of two numbers.

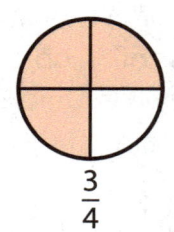

$\frac{3}{4}$

G

gallon (gal) a unit of liquid volume in the customary system. One gallon is equal to 4 quarts.

gram (g) a unit of mass in the metric system. A paper clip has a mass of about 1 gram. 1,000 grams is equivalent to 1 kilogram.

greater than symbol (>) a symbol used to compare two numbers when the first has greater value than the second.

H

hexagon a polygon with exactly six sides and six angles.

hierarchy a ranking of categories based on properties.

hour (h) a unit of time. One hour is equivalent to 60 minutes.

hundredths the parts formed when a whole is divided into 100 equal parts.

I

inch (in.) a unit of length in the customary system. A quarter is about 1 inch across. Twelve inches is equivalent to 1 foot.

inverse operations operations that undo each other. For example, addition and subtraction are inverse operations, and multiplication and division are inverse operations.

isosceles triangle a triangle that has at least two sides with the same length.

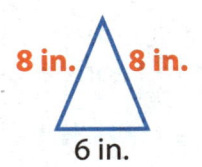

8 in. 8 in.
6 in.

K

kilogram (kg) a unit of mass in the metric system. One kilogram is to equivalent to 1,000 grams.

kilometer (km) a unit of length in the metric system. One kilometer is equivalent to 1,000 meters.

Glossary

L

length a measurement that tells the distance from one point to another, or how long something is.

less than symbol (<) a symbol used to compare two numbers when the first has less value than the second.

line (in geometry) a straight row of points that goes on forever in both directions.

line of symmetry a line that divides a shape into two mirror images

line graph a data display that shows data points connected by line segments.

line segment a straight row of points that starts at one point and ends at another point, or, a part of a line.

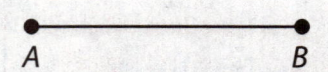

liquid volume the amount of space a liquid takes up.

liter (L) a unit of liquid volume in the metric system. One liter is equivalent to 1,000 milliliters.

M

mass the amount of matter in an object. Measuring the mass of an object is one way to measure how heavy it is. Units of mass include the gram and kilogram.

meter (m) a unit of length in the metric system. One meter is equivalent to 100 centimeters.

metric system the measurement system that measures length based on meters, liquid volume based on liters, and mass based on grams.

mile (mi) a unit of length in the customary system. One mile is equivalent to 5,280 feet.

milliliter (ml) a unit of liquid volume in the metric system. 1,000 milliliters is equivalent to 1 liter.

minute (min) a unit of time equivalent to 60 seconds.

mixed number a number with a whole part and a fractional part.

$4\frac{1}{2}$

Glossary

multiple the product of a number and any other whole number. For example, 4, 8, 12, 16, and so on, are multiples of 4.

multiplication an operation used to find the total number of items in equal-sized groups.

multiplicative comparison a comparison that tells how many times as many. For example, 7 × 3 = 21 means 21 is 3 times as many as 7, and 21 is 7 times as many as 3.

multiply to find the total number of items in equal-sized groups.

N

numerator the number above the line in a fraction. It tells how many equal parts are described.

$$\frac{2}{5} \leftarrow \text{numerator}$$

O

obtuse angle an angle that measures more than 90° but less than 180°.

obtuse triangle a triangle that has one obtuse angle.

odd number a whole number that ends in the digit 1, 3, 5, 7, or 9. Odd numbers do not have 2 as a factor.

operation a mathematical action such as addition, subtraction, multiplication, and division.

ordered pair a pair of numbers, (x, y), that describes the location of a point on a coordinate plane. The first number tells the point's horizontal distance from the origin, and the second number tells the point's vertical distance from the origin.

origin the point (0, 0) where the x-axis and the y-axis intersect on a coordinate plane.

ounce (oz) a unit of weight in the customary system. A slice of bread weighs about one ounce. Sixteen ounces are equivalent to 1 pound.

P

PM the time from noon until midnight.

parallel lines two or more lines that are always the same distance apart and will never cross.

parallelogram a quadrilateral with opposite sides parallel and equal in length.

parentheses ( ) a grouping symbol; they group parts of an expression that should be evaluated before others.

Glossary

partial products a strategy used to multiply multi-digit numbers. The products you get in each step are called "partial products".

$$\begin{array}{r}218\\\times6\\\hline 48\end{array}$$ (6 × 8 ones)
60 (6 × 1 ten)
1200 (6 × 2 hundreds)
1308

partial quotients a strategy used to divide multi-digit numbers. The quotients you get in each step are called "partial quotients".

pattern a series of numbers or shapes that follow a rule to repeat or change.

pentagon a polygon with exactly five sides and five angles.

perimeter the distance around a two-dimensional shape. The perimeter is equal to the sum of the lengths of the sides.

period a group of three related place values, usually separated by commas. Examples are the ones period, the thousands period, and the millions period.

Thousands Period			Ones Period		
Hundred Thousands	Ten Thousands	Thousands	Hundreds	Tens	Ones
4	6	7	8	8	2

perpendicular lines two lines that meet to form a right angle, or a 90° angle.

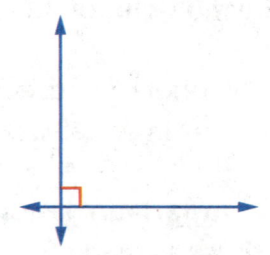

pint (pt) a unit of liquid volume in the customary system. One pint is equivalent to 2 cups.

place value the value of a digit based on its position in a number. For example, the 2 in 324 is in the tens place and has a value of 2 tens, or twenty.

plane figure a two-dimensional figure, such as a circle, triangle, or rectangle.

point a single location in space. Two lines cross at a point, and two sides of a triangle meet at a point.

polygon a two-dimensional closed figure made with three or more straight line segments that do not cross over each other.

Polygons	Not Polygons

pound (lb) a unit of weight in the customary system. One pound is equivalent to 16 ounces.

Glossary

prime number a whole number greater than 1 whose only factors are 1 and itself. For example, 2, 3, 5, 7, and 11 are prime numbers.

product the result of multiplication.

protractor a tool used to measure angles.

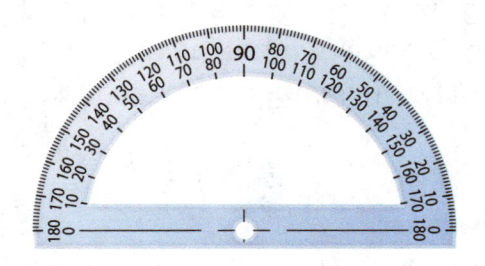

Q

quadrilateral a polygon with exactly four sides and four angles.

quart (qt) a unit of liquid volume in the customary system. One quart is equivalent to 4 cups.

quotient the result of division.

R

ray a straight row of points that starts at one point and goes on forever in one direction.

rectangle a parallelogram with four right angles. Opposite sides of a rectangle are the same length.

rectangular prism a solid figure with six rectangular faces.

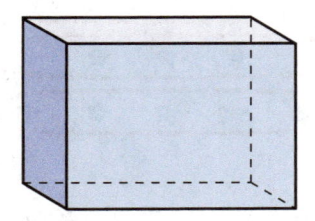

regroup to compose or decompose tens, hundreds, thousands, and so forth. For example, 10 tenths can be regrouped as 1 whole, or 1 tenth can be regrouped 10 hundredths.

remainder in division, the amount left over after equal groups have been made.

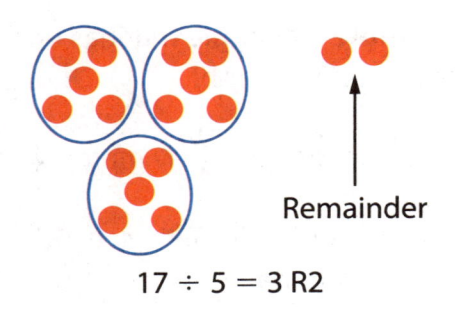

Remainder

$17 \div 5 = 3$ R2

rhombus a parallelogram with four equal sides.

right angle an angle that looks like a square corner and measures 90°.

right triangle a triangle with one right angle.

Glossary

round to find a number that is close in value to the given number by finding the nearest ten, hundred, or other place value.

row a side-to-side line of objects in an array.

rule in a pattern, a procedure that describes the relationship between one number or shape and the next.

S

scale on a graph, the difference between numbers labeling points on the axes.

scalene triangle a triangle that has no two sides with the same length.

scaling resizing a quantity by multiplying.

second (s) a unit of time. Sixty seconds is equivalent to 1 minute.

side one of the line segments that form two-dimensional figures.

solid figure a three-dimensional figure.

square a shape that has four sides of equal length and four right angles.

square unit a square with a side length of 1 unit that is used to measure area.

standard form the way a number is written using digits. For example, the standard form of *twelve* is 12.

sum the result of addition.

symbol any mark or drawing with a particular meaning, including numbers, letters, and operation signs. A symbol can be used to stand for an unknown number in an equation.

T

tenths the parts formed when a whole is divided into ten equal parts.

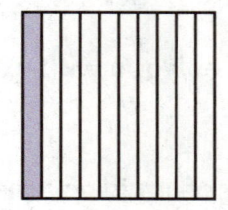

thousandths the parts formed when a whole is divided into one thousand equal parts.

three-dimensional solid, or having length, width and height. For example, cubes are three-dimensional.

trapezoid a type of quadrilateral. A trapezoid always has a pair of parallel sides.

Glossary

triangle a polygon with exactly three sides and three angles.

two-dimensional flat, or having measurement in two directions, like length and width. For example, a rectangle is two-dimensional.

U

unit fraction a fraction with a numerator of 1. Other fractions are built from unit fractions. For example, $\frac{1}{4}$ and $\frac{1}{10}$ are unit fractions.

unknown the piece or pieces of a problem that you are not given or do not know.

V

Venn diagram a drawing that shows how sets of numbers or objects compare.

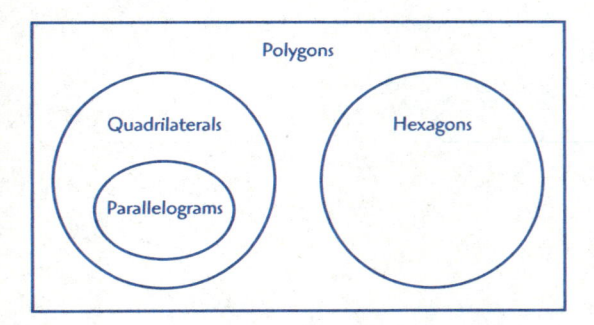

vertex the point where two rays or line segments meet to form an angle.

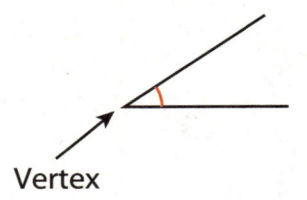

volume the amount of space inside a solid figure. Volume is measured in cubic units such as cubic inches.

W

weight the measurement that tells how heavy an object is. Units of weight include ounces and pounds.

word form (of a number) the way a number is written with words, or said aloud. For example, the word form of 105.7 is *one hundred five and seven tenths*.

X

x-axis the horizontal number line in the coordinate plane.

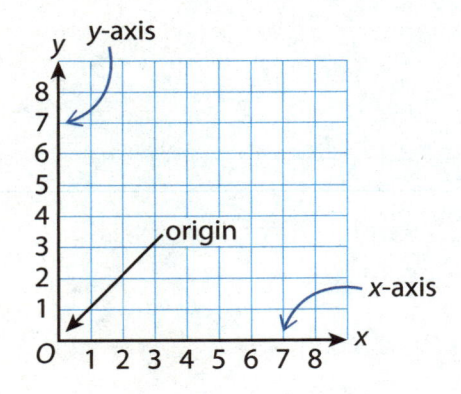

x-coordinate the first number in an ordered pair. It tells the point's horizontal distance from the origin

Glossary

Y

y-axis the vertical number line in the coordinate plane.

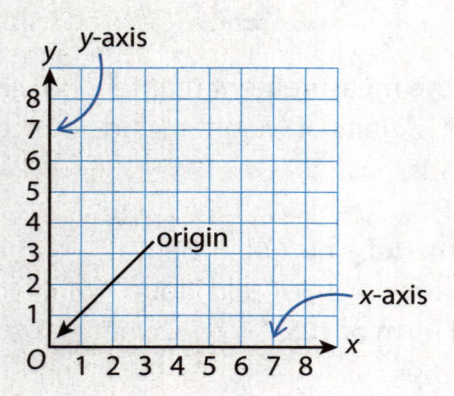

y-coordinate the second number in an ordered pair. It tells the point's vertical distance from the origin.

yard (yd) a unit of length in the U.S. customary system. One yard is equivalent to 3 feet.

North Carolina Standard Course of Study Coverage by Ready® Instruction

The chart below correlates the North Carolina Standard Course of Study to the *Ready® Instruction* lesson(s) that offer(s) comprehensive instruction on the standards. Use this chart to determine which lessons your students should complete based on their mastery of the standards.

North Carolina Standard Course of Study for Grade 5 Mathematics Standards	Ready® Instruction Lesson(s)
Operations and Algebraic Thinking	
Write and interpret numerical expressions.	
NC.5.OA.2 Write, explain, and evaluate numerical expressions involving the four operations to solve up to two-step problems. Include expressions involving: • Parentheses, using the order of operations. • Commutative, associative and distributive properties.	19, 20
Analyze patterns and relationships.	
NC.5.OA.3 Generate two numerical patterns using two given rules. • Identify apparent relationships between corresponding terms. • Form ordered pairs consisting of corresponding terms from the two patterns. • Graph the ordered pairs on a coordinate plane.	21
Number and Operations in Base Ten	
Understand the place value system.	
NC.5.NBT.1 Explain the patterns in the place value system from one million to the thousandths place. • Explain that in a multi-digit number, a digit in one place represents 10 times as much as it represents in the place to its right and $\frac{1}{10}$ of what it represents in the place to its left. • Explain patterns in products and quotients when numbers are multiplied by 1,000, 100, 10, 0.1, and 0.01 and/or divided by 10 and 100.	1, 2
NC.5.NBT.3 Read, write, and compare decimals to thousandths. • Write decimals using base-ten numerals, number names, and expanded form. • Compare two decimals to thousandths based on the value of the digits in each place, using >, =, and < symbols to record the results of comparisons.	3, 4
Perform operations with multi-digit whole numbers.	
NC.5.NBT.5 Demonstrate fluency with the multiplication of two whole numbers up to a three-digit number by a two-digit number using the standard algorithm.	5
NC.5.NBT.6 Find quotients with remainders when dividing whole numbers with up to four-digit dividends and two-digit divisors using rectangular arrays, area models, repeated subtraction, partial quotients, and/or the relationship between multiplication and division. Use models to make connections and develop the algorithm.	6

The Standards for Mathematical Practice are integrated throughout the instructional lessons.

North Carolina Standard Course of Study K-8 Mathematics © 2017. NC Department of Public Instruction. All rights reserved. Provided with permission from the Public Schools of North Carolina.

North Carolina Standard Course of Study for Grade 5 Mathematics Standards	Ready® Instruction Lesson(s)
Number and Operations in Base Ten continued	
Perform operations with decimals.	
NC.5.NBT.7 Compute and solve real-world problems with multi-digit whole numbers and decimal numbers. • Add and subtract decimals to thousandths using models, drawings or strategies based on place value. • Multiply decimals with a product to thousandths using models, drawings, or strategies based on place value. • Divide a whole number by a decimal and divide a decimal by a whole number, using repeated subtraction or area models. Decimals should be limited to hundredths. • Use estimation strategies to assess reasonableness of answers.	7, 8, 9
Number and Operations—Fractions	
Use equivalent fractions as a strategy to add and subtract fractions.	
NC.5.NF.1 Add and subtract fractions, including mixed numbers, with unlike denominators using related fractions: halves, fourths and eighths; thirds, sixths, and twelfths; fifths, tenths, and hundredths. • Use benchmark fractions and number sense of fractions to estimate mentally and assess the reasonableness of answers. • Solve one- and two-step word problems in context using area and length models to develop the algorithm. Represent the word problem in an equation.	10, 11
Apply and extend previous understandings of multiplication and division to multiply and divide fractions.	
NC.5.NF.3 Use fractions to model and solve division problems. • Interpret a fraction as an equal sharing context, where a quantity is divided into equal parts. • Model and interpret a fraction as the division of the numerator by the denominator. • Solve one-step word problems involving division of whole numbers leading to answers in the form of fractions and mixed numbers, with denominators of 2, 3, 4, 5, 6, 8, 10, and 12, using area, length, and set models or equations.	12
NC.5.NF.4 Apply and extend previous understandings of multiplication to multiply a fraction or whole number by a fraction, including mixed numbers. • Use area and length models to multiply two fractions, with the denominators 2, 3, 4. • Explain why multiplying a given number by a fraction greater than 1 results in a product greater than the given number and when multiplying a given number by a fraction less than 1 results in a product smaller than the given number. • Solve one-step word problems involving multiplication of fractions using models to develop the algorithm.	13, 14, 15, 16
NC.5.NF.7 Solve one-step word problems involving division of unit fractions by non-zero whole numbers and division of whole numbers by unit fractions using area and length models, and equations to represent the problem.	17, 18

The Standards for Mathematical Practice are integrated throughout the instructional lessons.

North Carolina Standard Course of Study for Grade 5 Mathematics Standards	Ready® Instruction Lesson(s)
Measurement and Data	
Convert like measurement units within a given measurement system.	
NC.5.MD.1 Given a conversion chart, use multiplicative reasoning to solve one-step conversion problems within a given measurement system.	22, 23
Represent and interpret data.	
NC.5.MD.2 Represent and interpret data. • Collect data by asking a question that yields data that changes over time. • Make and interpret a representation of data using a line graph. • Determine whether a survey question will yield categorical or numerical data, or data that changes over time.	24
Understand concepts of volume.	
NC.5.MD.4 Recognize volume as an attribute of solid figures and measure volume by counting unit cubes, using cubic centimeters, cubic inches, cubic feet, and improvised units.	25, 26
NC.5.MD.5 Relate volume to the operations of multiplication and addition. • Find the volume of a rectangular prism with whole-number side lengths by packing it with unit cubes, and show that the volume is the same as would be found by multiplying the edge lengths. • Build understanding of the volume formula for rectangular prisms with whole-number edge lengths in the context of solving problems. • Find volume of solid figures with one-digit dimensions composed of two non-overlapping rectangular prisms.	27, 28
Geometry	
Understand the coordinate plane.	
NC.5.G.1 Graph points in the first quadrant of a coordinate plane, and identify and interpret the *x* and *y* coordinates to solve problems.	29, 30
Classify quadrilaterals.	
NC.5.G.3 Classify quadrilaterals into categories based on their properties. • Explain that attributes belonging to a category of quadrilaterals also belong to all subcategories of that category. • Classify quadrilaterals in a hierarchy based on properties.	31, 32

The Standards for Mathematical Practice are integrated throughout the instructional lessons.

Acknowledgments

Illustration Credits

page 38: Fian Arroyo
page 56: Rob McClurkan
page 108: Fian Arroyo
page 230: Cool Vector Maker/Shutterstock (sandwich)
page 230: Macrovector/Shutterstock (pizza and hot dog)
All other illustrations by Sam Valentino.

Photography Credits

page 96: M88/Shutterstock (goat)
page 96: Tsekhmister/Shutterstock (pig)
page 98: 3drenderings/Shutterstock
page 99: Africa Studio/Shutterstock
page 182: Romas_Photo/Shutterstock
page 186: Maksym Bondarchuk/Shutterstock
page 188: LianeM/Shutterstock
page 189: Madlen/Shutterstock
page 224: Stephen Mcsweeny/Shutterstock
page 231: Zeljko Radojko/Shutterstock
page 298: Miro Novak/Shutterstock
page 302: Iryna Rasko/Shutterstock
page 304: stockcreations/Shutterstock

Background images used throughout lessons by Ortis/Shutterstock, irin-k/Shutterstock, and Kritsada Namborisut/Shutterstock.